Reinventing
the Forest Industry

Other Books by Jean Mater

Citizens Involved: Handle With Care!

The Public Acceptance Assessment

Public Hearings Procedures and Strategies:
A Guide to Influencing Public Decisions

Marketing Forest Products

Reinventing
the Forest Industry

Dr. Jean Mater

GreenTree Press
Wilsonville, Oregon

Cover Design by Catherine M. Mater and
Margaret Puckette

GreenTree Press
a division of
BookPartners, Inc.
P.O. Box 922
Wilsonville, Oregon 97070

Dedication

*This book is dedicated
to the memory of
Milton H. Mater
and to my family*

Table of Contents

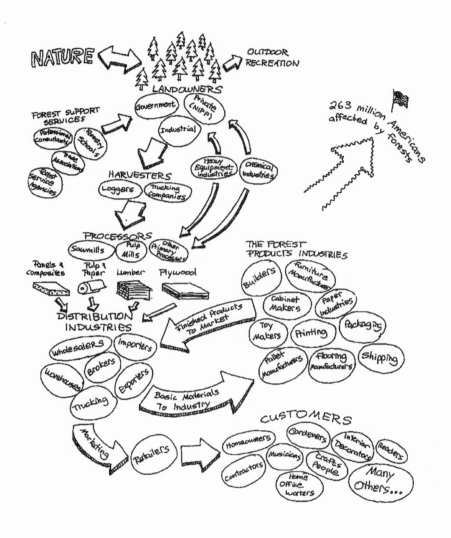

INTERDEPENDENCE of the FOREST PRODUCT INDUSTRIES

Preface

Every American has a stake in the forests that enrich our land. However, as stakeholders we act like haggling relatives fighting over our forest inheritance. Each stakeholder wants assurance that the will provides for his or her share.

Not so long ago the future distribution of forest wealth was relatively simple: provide for timber production and save some forests for beauty and recreation. Today, many stakeholders with other ideas are quarreling about the forests' legacy to future generations.

From California to Maine, Alaska to Hawaii, all Americans want something from the forests: Farmers and developers want land; homebuilders and homeowners need lumber from the timber; hikers, campers, and backpackers demand that the forest's enchantment be preserved; skiers and snowmobilers crave access to the forest; nature lovers yearn for a primeval forest to nurture their spirits and provide homes for birds, animals, and fish. Biologists and fish and game experts demand their share for endangered species. All claim they're entitled to the fresh air and water scrubbed clean by the forest.

In the midst of these quarrels are the more than twenty-five million people in the forest industry who earn their living planting, harvesting, and converting trees to lumber, furniture, and homes. They provide essential services to nurture these activities, struggle for enough timber to provide the housing, furnishings, cabinets, and other shelter comforts from the forests, and distribute wood products to consumers. Adding to the struggle are the fifty million Americans, who have joined organizations to preserve the environment, trying to keep the chainsaws from these forests. Silent, but more significant, are the more that 260 million Americans who live in households that enjoy the comforts provided by wood construction, wood furniture, wood floors, and other amenities.

As the human family grows, the relatives have become more quarrelsome about dividing the wealth; and, ultimately, the stakeholders have become antagonists all pulling at the forest industry.

The forest industry is caught in the tug-of-war among these competing stakeholders. Bitter disputes are disrupting entire communities and industries, endangering our children's and grandchildren's future. Emotional opinions range from, "We should never cut another tree," to, "What's wrong with the way the forest industry is managing now?"

We appear to be fighting over every tree. The federal government has sought unsuccessfully to find a balance between interests, but there is little unanimity within the forest industry or the preservation community on how to achieve peace in the forest. Incredible energy has been wasted on confrontations between stakeholders, each spending vast sums to gain power to enforce its view.

We need to divert our energies from confrontations to finding solutions to accommodate the needs of all stakeholders.

The forest industry now recognizes the power of the new environmental culture that restricts its former free-wheeling activities. Environmentalists are just beginning to understand that loggers, sawmill workers and paper mill-workers need jobs to support their families.

Reinventing the Forest Industry outlines a solution: to become part of the current environmental culture, the forest industry must substitute new practices for many of the traditional methods once acceptable in another era. Environmental myths assume that the forest industry ruins whatever it touches. The old "cut-and-run" reputation persists, and the public does not trust the forest industry's environmental standards. Unfortunately, the industry and the country pay dearly for this lack of trust. With scientific and responsive forest management so essential for today and the future, the public also suffers from that negative perception of the forest industry.

The forest industry will regain the opportunity to be stewards of the forests only if it wins the public's confidence. Education and public relations have failed in this mission. Regaining stature will come through *Reinventing the Forest Industry*. When the forest industry is trusted not to "spoil" or "devastate" forests, it will put a new spin on the image of the forest industry.

Reinventing is a plan, based on extending new practices already under way and eliminating actions that disturb the public. It is predicated on changing the forest industry, not the public. Gaining the public trust should minimize conflict and shift technical forest management from the courts back to qualified scientists.

Forests take years to grow. If we do not plan now for future demands from the forests, how will we avoid chaos and hardships in the next century, when our population doubles again?

As the principal grower of forests, the forest industry is the key to preserving our forest wealth. A fuller understanding of the forest paradoxes, the environmental movement, the reasons for the negative public image of the forest industry, and the attempts by the industry to cure the sick image demands that the industry reinvent itself to better serve as the steward of the forest.

The conclusions in this book are mine, but many others have contributed valuable insights. Thanks to DaNelda Strode, Margaret Puckette, and the Montana group — David Spencer, Willow Creek Wood Productivity Training Center; Harry Urban, *Wood and Wood Products;* Steve Ehle, *Wood Digest;* Ed Jerger, Wood Production Consultant; Jim Poxleitner, Montana Power Company; and Judy Riggs, for their assistance and input. To my many associates throughout the United States and Canada who shared their thinking with me, please know that I am grateful.

I cannot adequately express my appreciation to Michael Jenkins and the Catherine and John T. MacArthur Foundation for the grant that enabled and encouraged me to write this book. It was joyous labor that I hope provides readers insights into the magnitude of the problems and possibilities for solution.

Jean Mater
Corvallis, Oregon
1996

Introduction

Residents of the small communities that dot the Northern California Sierra Nevada Mountains worry as they watch many of the spectacular fir and pine trees in the surrounding forests dying of disease and old age. A careless match could ignite a disastrous fire in the dry forest, devouring homes and threatening lives. The stench of acrid smoke and the memories of last summer's fire in a neighboring forest linger in their nightmares. They are afraid, and they don't know what to do.

However, these people find they have little say about the potential danger facing them. The thousands of city folk who visit the mountains each year to hike and camp in those old forests claim the forest as theirs. And, because of this, they fight the loggers' chainsaws in their forests.

While the campers and residents debate, nothing is done to prevent the looming catastrophe. The leader of one of the imperiled communities shakes his head mournfully.

"Soon neither side will have anything to fight over," he says. "A fire will settle the argument."

Meanwhile, three thousand miles east, communities in Maine and northern New England are fighting desperately to shield their forests from the timber and paper industry, and a horde of other enemies: tourists, fishermen, hunters, hikers who tramp into the forests, and the refugees from the cities who build homes where trees used to stand. Some 15,000 members of a group called Restore: The North Woods petition to preserve those forests in a new national park. Hunters and fishermen in the Sportsman's Alliance of Maine bitterly oppose the park. Passions run strong; hostility invades the once-peaceful region.

The forest community in Maine is fighting an uphill battle against "a loaded gun pointed at the head of the Maine economy." At least that's the way the state's governor, Angus S. King Jurl, described the "clearcutting referendum" that was on the ballot for the fall election.

"It's been sold as a referendum to ban clearcutting, but there's only one sentence in it that mentions clearcutting," says John McNulty, vice-president of the Bangor-based Seven Islands Land Company, a family-owned enterprise that manages approximately one million acres of forest-land.... Our estimation is that it will reduce harvesting in the state of Maine by as much as 50 percent. The way I see it, that means for 40,000 people in Maine who work in the forest products industry, they'll have a 50 percent chance of getting a job."

The proposed ban would have covered the 10.5 million acres under the control of the Maine Land Use Regulation Commission.... It addressed clear-cutting by limiting harvesting to no more than one-third of the marketable trees on each acre in a fifteen-year period. Also, logging operations would be barred from opening a forest canopy of more than half an acre, except along roads.... (Jeff Ghannam, *The Forestry Source*, Society of American Foresters, June 1996)

CONFRONTATIONS IN THE FOREST ARE SYMPTOMS OF PEOPLE VERSUS PEOPLE

It would be shortsighted to downplay these antagonisms as more disputes between the forest industry and environmental advocates for control of the forests. The confrontations characterize tensions between developers and forests over land that is becoming scarce; between sawmills and pulp mills, and clean air and water; between campers who nestle under the trees and loggers who harvest the trees; and between jobs for people and protecting endangered species.

The striving to keep the values we call "quality of life" while providing jobs for a fast-growing population demanding more jobs and more living space is swallowing more of our time and money.

Communities over most of the United States are struggling to accommodate more people within previously determined "growth boundaries." There is more traffic on overcrowded roads, more houses and businesses, more ski lifts for skiers, and a call for more pristine areas for backpackers. It is handier to blame the automobile, forest industries, miners, and "greedy" developers for the inconveniences of modern life than to face the realities caused by an expanding population.

The forest industry is often the villain in the everyday dramas of people versus people. It has become the victim of difficult policy options that affect everyone. They face choices to harvest trees to provide homes for humans or not harvest to provide habitat for wildlife. Should the forest industry set aside forests for hunters and backpackers, who come down hardest on those who depend on the forests for their income?

Meanwhile, the forest industry wrestles with the public's expectation that all forests can fill all their desires at the same time. To satisfy incompatible demands, government policy applies band-aids to regulate the forests, even those privately owned, in an attempt to please everyone.

A made-to-order emblem of our growing frustration with today's world, forests symbolize America's love affair with nature. To nature advocates, the logger's chainsaw is a dagger piercing the heart of the forest. Popular opinion views environmentalists as more interested in the forests, and industry as more interested in the trees.

It is environmentally chic to suggest that we can save trees by substituting steel, cement, or plastics for wood. That these substitutes are made from carbon, petroleum, silica, and metals mined or extracted from the earth appears to carry little weight. The public doesn't want mines, oil fields, or chemical plants either. Extracting these materials is hardly more attractive than clearcut forests that are the bone of so much contention — and other resources do not renew themselves, as the forests do.

Recently, David Todd of Shrewbury, Massachusetts, saw in the men's room of a public library in Fairbanks, Alaska, an electrically powered hand-drying device that bore the message:
"Dryers help protect the environment.
They save trees from being used for paper towels.

- - -- -- -- -- -- -- -- -- -- -- -- -- -- -- -- --

They eliminate paper towel waste.

They are more sanitary to use than paper and help maintain cleaner facilities."

On a little checking, he found that electricity for the dryer came from a coal-powered generating plant in Fairbanks. The coal is obtained by surface mining at Healey, about eighty miles south of the city, and is hauled to the generating plant by a diesel-powered train. Todd thinks the dryer no doubt saves the janitor a lot of work picking up paper towels, but he's not so sure about the environment. *(Chemical and Engineering News, August 26, 1996).*

WRONG SOLUTIONS DON'T SOLVE REAL PROBLEMS

Adding up all these factors and factions, it's obvious that the imperative to apply our energies to discover how to reconcile the competing demands has never been so urgent. The world's resource problems are not likely to be solved by merely keeping chainsaws out of forests, or reducing our consumption of the world's goods and making do with less. The public hasn't bought the idea of "doing without" to save the world. Doing without also runs contrary to current government policies that depend on increasing consumption to provide employment and taxes to fund public benefits.

In the past decades environmental advocates have succeeded in changing Americans' attitudes about our relationship to the natural world. Environmental awareness is thoroughly implanted into public opinion. Americans under forty believe that environmental concerns are givens, not issues. In other words, the transition from the pre-environmental paradigm has already happened. Public opinion, in turn, is persuading the forest industry to adopt the principle of caring for the environment. Paying attention to the environment has become a good thing to do.

The environmental consciousness-raising has penetrated deeply into all industries. In fact, the forest industry has embraced so many of the environmental principles that some environmentalists now feel deprived of their cause. Dr. Patrick Moore, one of the original organizers of the activist group Greenpeace observed that eco-extremists are pressing for a new cause, now that the forest industry has adopted many environmental tenets.

UNTOUCHED NATURE OR SOPHISTICATED CIVILIZATION?

Both the forest industry and the environmental interests have grown more pragmatic about America's concerns for environmental values. Extreme positions win media attention, but the mainstream concerns on both sides are inching toward common goals. The forest industry is turning a dizzy 180 degrees in its environmental awareness, embracing the concept of sustainable forest management. The environmental community is making a ninety-degree turn, recognizing that it should be courting loggers, sawmill workers, and forest-dependent communities to become their allies instead of dislocated victims of environmental awareness.

The forest industry has already taken the first tentative steps to reinvent itself to match the world's new earth-consciousness. It has walked the first mile to peace in the forest with its commitment to sustainable forestry. It still faces the challenge of walking that last mile and acknowledging as well a social responsibility as steward of the forest.

The forest industry's and environmental advocates' positions are now closer than either recognize. A quick survey of the forest industry professional and trade journals *Forest ipl wrong* reveals how industry practices are changing in response to our environmental culture. The Forest Products Society Technical Group on Wood and the Environment is a only a few years old. The Society of American Foresters is committed to ecosystem management, which has dominated forestry discussions in the nineties. Ecological properties of forests and integrating of societal values have become respected parameters in forestry, though the discussion is still inner directed, using fairly technical language. Labels like "eco-extremists" and "enviros" still appear in industry trade journals.

A similar quick scan of the publications of the environmental groups — Sierra Club, National Wildlife *Enviros wrong* Federation, Audubon Society, Izaak Walton League, and others — reveals that they usually omit discussion about the turnaround of the forest industry. When mentioned, the industry is still pinned with the "greedy corporation" and "profit-monger" labels. Thus, the public is unaware of the changes in the forest industry and is still influenced by the cut-and-run image of yesterday.

Strangely, the forest industry itself knows little more about the new environmental practices than the general public. The industry has yet to realize the broad span of how wood is involved in American life. Pallet, cabinet, and furniture manufacturers, for example, often know as little about forestry practices as foresters know about the use and requirements of the products of the forest.

A whopping 81 percent of the companies in the forest industry are small establishments with fewer than twenty employees. Most of the forest land in the country is owned by 10,000 small, non-industrial private forest owners (NIPFs) whose holdings differ widely from state to state. Indiana NIPFs own thousands of small, isolated acres of hardwoods, located mostly on farms. South Carolina's NIPFs often contain commercial conifer stands, while Utah's private forests are most often located on ranches or vacation properties that are rarely managed for timber. Few in the forest industry qualify as the legendary "greedy, profit-monger" timber barons. Facts, however interesting, appear irrelevant in the emotional climate stirred by concern over the environment.

As we turn to the next century, the forest industry is still experimenting with changing its image from a symbol of man against nature to an industry that is working with nature. As more people press against the boundaries of nature, the forest industry has the opportunity to demonstrate how to operate between the extremes of untouched nature and sophisticated civilization.

Chapter 1

A Frenzy of Change

"Some kids are born with a silver spoon in their mouth. I was born with a log," says Frank Simmons Jr., president of Simmons Lumber Company of Cooke City, Idaho. Frank knew before he entered kindergarten that he would someday run a sawmill. His father, Frank Sr., had invested his savings in a sawmill, a small operation just after World War II, when anyone with a few bucks and a lot of knowhow could make money turning out lumber. The GIs were returning home, eager to start families and build the homes they dreamed about during the long war years. Frank Sr. picked up some timber contracts, and all he had to do was work twelve-hour days seven days a week, and he could make a profit.

Unlike his dad, who had dropped out of high school to join the army, Frank Jr. had the money to go to college and major in forestry, earning a Bachelor of Science in Forestry. His specialties were Forest Engineering, Forest Management, Forest Products, and Forest Recreation Resources.

His dad proudly called him his "college boy." Frank Jr. pulled green chain in the mill during summers and holidays, learned the mill lingo, grew proper calluses on his hands, and traveled with his dad to learn new ideas from other sawmills. He loved the pungent smell of sawdust and freshly sawed logs.

The Simmons mill did make money. Frank Sr. spent a lot of the profits on the latest equipment. Frank Jr.'s mother accused him of caring about the mill more than her. Somehow she was able to save enough money to buy good furniture for the home they had built in the late 1950s, after one of the prosperous years. When Frank Sr.'s health began to fail, Junior eagerly jumped in to manage the mill and take over the responsibility to keep sixty-five employees paid every week. Frank spent his days negotiating for timber, battling economic cycles, and selling lumber to a broader market to keep the men working steadily.

That was how Frank Jr. expected to live the rest of his life. Then, somewhere in the early 1980s, his work schedules changed. Days were spent poring over the new "regs" about harvesting and OSHA (Occupational Safety and Health Administration) safety rules, and attending meetings about threatened closures of his timber supply. As the 1980s waned, he was spending so much time at emergency meetings and learning new regulations that he sighed, "I should have majored in political science. It doesn't do me much good to be a forester nowadays."

AN EPIDEMIC OF CHANGE

Frank Jr. found himself caught in an epidemic of rapid change that was transforming his life and the way he conducted his business. Neither education nor experience prepared him for the changing world that Peter Drucker, the guru of modern management, summed up by saying, "No century in recorded history has experienced so many social transformations and such radical ones as the twentieth century." Drucker is a prominent management consultant and former Professor of Management at New York University's Graduate School of Business, author of *The Effective Executive, The Age of Discontinuity, The Practice of Management,* and other pathbreaking books on business management.

Simmons says "amen" to that. As the twentieth century ends, media headlines and statistics confirm what he's known in his bones — that the world is changing too fast for comfort. Like others nearing middle age, his expectations for the remainder of his life differ from his parents'. He planned to work in forestry the rest of his life. He worries that he may have to change careers. He thought his college degree guaranteed security. His kids are learning that a diploma is only the beginning of a lifelong education process of swimming with the current of change.

The *Wall Street Journal* warns Simmons: "It's time you became a manager of change." *U.S. News and World Report* tells him: "The Internet will change everything — and everyone wants a piece of the action." He is unsettled by a *Business Week* report that says that fourteen of the forty-three companies considered the "best run" in America fell into financial trouble because they failed to react and respond to change.

He is not alone when it comes to grappling with change. Industries throughout America are warned that they are the victims of change. Like Frank Jr. they are finding that the notion of coping with change, of being the victim of change, is practically obsolete. Instead, industries, schools, governments, churches, and other institutions are looking for a better way to handle change rather than merely coping. Instead of contending with change, they are learning how to use it to grow.

Simmons is trying to understand how to use these changes to keep his sawmill running. He sees industries taking on a new role in society — transforming the manufacture of chemicals, the building of houses, the education of both the young and the mature, the practice of law and medicine, and governments at every level.

A new role in society is a heavy load for a fellow who was born to run a sawmill and would rather just keep his sixty-five employees in jobs.

Simmons puzzles about how the forest industry suddenly seems to be a symbol for ruining the planet. None of his forestry courses in college prepared him to be a villain. He is well aware that the survival of his mill depends on the availability of forests. He knows that humans depend on forests for survival but has difficulty swallowing the "tree huggers'" notion that his mill affects the air they breathe and the water they drink thousands of miles away. But he recognizes that people believe they are entitled to clean air and water nowadays, and he has a responsibility in providing them.

The forest industry is so pervasive in everything we do that most people, including those in the industry itself, fail to realize its significance to life. Simmons thinks "those environmentalists" make too much of the mystical beauty of the forest. "Like a religion," he complains, resenting those who don't know anything about managing a forest depriving him and his crew of the logs they need.

NAFTA
1994

~~1994~~ Post Nafta?

Depending on a highly visible natural resource that provides so many benefits, Simmons finds the timber he's hungry for appears low on the public priority list. The demands of ardent advocates with enough causes to fill a book seem to constantly thwart the forest industry's timber appetite.

Until the late twentieth century, the industry could fill its shopping basket from ample supplies of timber. Today supplies are still ample, but timber is no longer king of the forest, even in the regions where the industry still exerts a royal privilege to cut whatever timber it needs.

FORESTS AT THE EPICENTER

The industry is torn by changes that are forcing it to fundamentally reinvent itself in order to survive and compete. Like an earthquake that shakes a houseful of furniture, breaking glassware, tumbling books from shelves, the changes are shaking up the forest industry. The forest industry is at the epicenter of the quake, forced into moving from its insular view as a provider of goods, jobs, and income to a position of responsibility for the future of the earth.

Simmons finds himself constantly pushed to defend his position. His son, with the assurance of a high school sophomore, accuses him of "raping the earth" and killing the planet by cutting trees. "Where did you get such an idea?" Frank fumes. He's furious when his kids talk about greedy lumbermen and money-hungry businessmen.

A colleague from a nearby papermill that buys the chips produced from the residues of Simmons's mill reports the same experience when he was invited to his daughter's fifth-grade class to talk about making paper. He planned to describe how much paper his plant was producing, how many jobs they provided, and how trees were used for fiber. He never got the chance to give his little speech. The class knew what they wanted to hear, and they were the attorneys, the judge, and the jury.

"Why are you cutting down the rainforest to make paper?" one asked. The class nodded in unison, glaring accusingly.

Another hand waved insistently: "Is it true that it takes a whole forest to make the Sunday paper? How could you cut down all that old growth?"

"Why do you throw all that dioxin in the river?" a third shouted. "Doesn't it give people and fish cancer?"

His forestry work spills over into his personal life, and Simmons finds conversation with one of his neighbors irritating. His neighbor likes to taunt him, calling him "the timber baron," and rattles off statistics about overcutting the forest and destruction of the ozone layer and the greenhouse effect. Frank shows him the book by Douglas W. MacCleery of the U.S. Forest Service about the state of American forests. "See," Frank explains, "We have built millions of homes, but America has about the same forest area as it did in 1920. The forest is growing more than a third faster than it's being cut."

"Hmph," replies his neighbor. "What do you expect the Forest Service to say? They're in bed with the timber barons."

And still the aftershocks from the quake spread. At their neighborhood barbecues they've added "no talk about logging" to the taboo topics of religion and politics because the arguments became too heated. The neat notions Simmons learned about the way land is allocated and the comfortable Multiple Use Model that Congress created in the Multiple-Use Sustained Yield Act of 1960 no longer satisfy the public. Many constituencies were feeling that managing national forests for outdoor recreation, wildlife, grazing, and watershed protection, as well as timber, did not go far enough in protecting the forests.

His former pat answer to people worried about a timber famine was that we can prove scientifically that we have been growing trees faster than we cut them. Now no one believes him.

THE OLD ANSWERS DON'T SATISFY ANYMORE

The myth about running out of trees has been extensively played out in the media over the past two decades. In actuality, we have more trees now than in 1970, the year of the very first Earth Day. For example, states such as Vermont, Massachusetts and Connecticut were about 35 percent forested, and today they have increased to 59 percent. Even though the percentage has increased, forest growth has exceeded harvests since the 1940s. Through proper land management and by providing conservation awareness, we will have plenty of trees to use in the next century and beyond.

To people worrying about wildlife habitat and survival, he recited the respectable current count of deer and elk. They challenge that statistic.

One of his wife's friends baits him every chance she gets, claiming that the timber beasts are cutting too many trees just to manufacture paper. She's not impressed when he tells her that paper now uses almost up to 50 percent recycled fiber. "Yeah, from clearcuts of old growth," she retorts.

He boasts about how the forest industry is now saving trees by using parts of the tree that they used to waste to make wood panels out of chips and glue. "The boards we make are often better than the ones nature makes," he says. "They use formaldehyde, and anyway, it's not natural," a friend replies.

He tries to explain the new sustainable forestry and ecology to a fishing companion. His friend expresses skepticism about certifying a sustainable forest.

The old pat answers don't seem to satisfy anyone anymore, though he feels satisfied that reserving some forest-land for animals, wilderness, spotted owls, and unique characteristics is pretty good. Even his minister protests during a golf round that these are not solving the real problems. "Window dressing," the minister complains. "Where are we going to put that new low-cost housing development?"

"Don't blame that on the forest industry," Simmons retorts. "We just have too many people."

Remembering his conversations with friends and acquaintances, he says, "I should have taken up debating. That's about all I do nowadays."

As a sawmill owner responsible for the livelihood of sixty-five men and the income of many stores and services in the community, Simmons reacts by resisting the new environmental ideas that almost all his acquaintances seem to have adopted.

He surveys the forests surrounding his mill and shakes his head in dismay that he may have to close the mill — put his people out of work — because he can't harvest the timber. "How did we get into this mess?"

People who care about the environment shake their heads at clearcuts and ravaged forests and also ask, "How did we get into this mess?"

‑‑‑

One answer to that frustrating question lies in recognizing changes that are battering the forest industry because it occupies valuable space and provides essentials to the growing population.

Five changes help explain how we got into this mess:

1. More Americans vying for the space now occupied *True* by forests.
2. The legitimizing and empowering of citizen participation in government decisions. *Fed by 5*
3. The increase in the number of people who believe *Fed by 5* they are rightful stakeholders in the industry.
4. Complying with the new environmental values as a new litmus test for consumer acceptance. *True*
5. The growing public demand for more and broader social responsiveness from the forest industry. *5, True.*

These changes together helped create the "mess" and are transforming the forest industry. They throw light on why the recreation needs of a more urban population confront the industry at almost every turn. Clearcuts have never been popular, but now, with more people enjoying recreation and tourism in the forest, they symbolize what the public dislikes most about the forest industry. The barren clearcuts throw a mantle of suspicion, and this negative image has been laid on the back of the forest industry. It has proven a heavy load.

Population Shifts Pose New Challenges
WASHINGTON — Don't look now. The country's changing again.

Seismic shifts are occurring in the character of the nation. People are moving, and not in predictable ways....

But deep in employment figures are signs of a major transition in the way Americans work and where they live.

And increasingly in the next several years a new buzzword will muscle itself into the national dialogue: exurbanization.

It means that the new growth areas of the country aren't the central cities, nor their suburbs, nor creeping areas of settlement beyond the suburbs.... The growth will be in small communities beyond the current population areas — farm regions, pockets where high school graduation rates are high, places where taxes are low.

"This," said Martha Farnsworth Riche, director of policy studies at the Population Reference Bureau, "is the biggest unnoticed trend in the country."

It is, however, beginning to be noticed. A Georgia Tech study team headed by Arthur C. Nelson found that nearly three out of every five jobs created between 1963 and 1987 were located in exurbia....

Meanwhile, however, plain old economic forces have been at work, and light and high-tech manufacturing plants are relocating to places like south-central New Hampshire, a swath of land between Topeka, Kansas, and Kansas City; and the western counties of Wisconsin. Much of the relocation is also occurring in the Northern Plains and the Southwest. (David Sherman, *Boston Globe,* April 11, 1994)

THE UNDERLYING PROBLEM OF MORE PEOPLE

The chaos in the forest industry house left by the transforming earthquake proved to be minor compared to the cracks appearing in the basic foundation of life on earth. It's a crack too big for Simmons to even contemplate.

Much of the friction with the environmental groups — a catch-all term for the groups that have made life more difficult for the industry — stems from the unprecedented expansion of the world's population.

"Bunk!" Simmons protests. "Don't give me that. I've heard it before. It's just an excuse for more regulation."

And he has heard it before, from Arthur Godfrey, the popular television pundit who announced in 1970: "It is predicted, scientifically, that we shall export our last load of wheat in 1976, just six years away! Why? Because we will have no more surpluses to export. In fact, we won't have enough for ourselves. If the population of the world doubles in thirty years, that means it will double here too...."

He heard it in the predictions of ecologist Paul Ehrlich who, in 1970, warned that "our fragile planet has filled with people at an incredible rate."

Simmons hasn't seen the numbers. That's not the stuff he reads. However, statistics show that, within his lifetime, since 1940, the population of the United States has doubled, from almost 132 million to 263 million in 1996. In other words, everyone in the world who has reached the age of forty has seen the population double in his lifetime. Some demographers estimate that the earth can support a maximum of one billion to one trillion people. We are already at six billion, and in a little more than fifty years the population is projected to double again.

True, we do not lack food. We're not elbowing each other for more room yet, but accommodating more people on the planet is one of our harsh realities. These accommodations add to life's irritations and provide the unrecognized and rarely discussed agenda behind many regulations passed under the aegis of protecting the environment.

When it takes thirty minutes to drive across town instead of the previous fifteen-minute trip; when old-timers in town who have been satisfied with their septic tanks are taxed to pay for a sewer system to meet the needs of newcomers; when proposed housing developments breed controversy; when urbanites insist on protecting their recreation areas by keeping the loggers out of forests, we are experiencing the first inconveniences brought about by a growing population.

Stories in the local newspapers warn of impending disaster:

Simmons's mill has been running on a thread of logs. An endangered bird has been discovered in Frank's forest: Logging will stop. The consequences: Frank is making a deal for Russian and New Zealand logs.

The town has organized an economic development commission to attract some industry, any kind of industry, to provide jobs. The "we need more jobs" faction opposes the new residents and retirees who rebel at community growth. Destroyed jobs and more expensive wood for housing have lost their influence on public decisions.

This dispute is not unique to Simmons's town. It is happening all over the United States.

> In 1995 a Canadian natural resource company proposed a development to mine $750 million worth of gold and silver out of a peak in Montana. Backers of the venture promised jobs. Environmentalists cried, "Foul!" The growing recreational tourism business protested, "A mine will kill us." When the *Billings* [Montana] *Gazette* asked residents of the region how they felt about the mine, almost a third feared too many people would come, and close to half frankly thought that Montana, with its 850,000 population, was already approaching its growth limits. Tourists crowd the area during the summer and fall. They bring money and tourism-related jobs, but they don't stay. The comments are reminiscent of Oregon's (former) Governor Tom McCall's historic remark to visitors in the 1970s, "You're welcome to visit, but please don't stay."

The threat of an expanding population helped spark the environmental organizations to action in the 1960s and 1970s. The Audubon Society, the Izaak Walton League, and other environmental organizations launched campaigns warning about the population expansion. They fear that the increasing population combined with American extravagant consumer habits will rob the earth's resources and despoil the environment.

Forests are the first victims of our frustration with the prospect of enjoying less of the earth's bounty in the future than we have come to expect as our right in the twentieth century. We are facing the dilemma of destroying forests by cutting either too many trees or too few trees.

The Gypsy moth has been allowed to freely attack the Southeastern hardwood industry. One forest in Virginia has 90,000 acres of dead timber, killed by the Gypsy moth. The current environmentally acceptable forest management practice of avoiding chemical applications in the forest has allowed the Gypsy moth to run rampant through the mature oak, making it more susceptible to defoliation and disease. The infestation and disease might have been controlled using modern aggressive, scientific management principles, including harvesting the dead and dying trees. Under the present rules only timber that is dead may be harvested. Dead timber has no value for the timber industry, but environmental scientists consider it valuable in ecological management. Diseased mature trees may yield good lumber, but they are not allowed to be harvested until they are dead, when they may be worth less as lumber.

Frank's town, Cooke City; and the Southeastern hardwood industry were squeezed in the environmental movement's effort to raise America's consciousness about the environment. The successful effort ran roughshod over natural resource-based industries. Now the thirty-year-old environmental movement is maturing. On the twenty-fifth anniversary of the celebrated Earth Day, the National Audubon Society's Brock Evans stated: "We were on a crusade for righteousness and good, out to save the world. Our intentions were high. We felt we were doing the right thing, but we were insensitive. All these people [farmers, landowners, timber workers, municipal officials, and others] could have been our allies, but we ignored them. Now we are reaping the bitter fruits."

CITIZENS INVOLVED — WATCH OUT!

Promoted originally to assure a role in public decisions, citizen participation has made its way as a force in the board rooms of industry. In the United States, citizens have long talked back to their government and have rarely been reticent about expressing opinions on all matters. The "Great Society War on Poverty" programs originated during the presidency of Lyndon Johnson in the 1960s institution-alized and mandated that citizens participate in the Housing Education Welfare (HEW), Housing and Urban Development (HUD), and the Community Services Administration (CSA) Federal Grants-in-Aid programs on housing, public health, and safety. In the 1970s, workshops and courses blossomed to empower citizens to speak up and demand their rights. The forest industry paid little attention to the growing vigilance of citizens with complaints. The citizens were active in government social issues and, for a while, business felt comfortably secluded from the vigorous citizen groups.

By the 1970s, however, the cocoon that sheltered industry from citizens' barbs was pierced by the volume of new legislation that called for citizen participation in decisions. Matters that before were the exclusive province of business became the public's business. The forest industry suddenly found itself on the firing line. It was unprepared to deal with citizens who demanded action in their forests that were often at variance with standard forestry practices.

The Coastal Zone Management Act of 1972, the Resource Conservation and Recovery Act (RCRA) of 1976, and the Federal Water Pollution Control Act Amendments of 1972 required participation by citizens. In the period between 1970 and 1974 alone, five Environmental Protection Agency (EPA) grants set rules for citizen participation. By 1978, the total was eight.

The forest industry, reeling from the sudden requirement to devote time and personnel to handling meetings with the public, could not anticipate the far-reaching impact of these citizen participation mandates. Foresters found themselves badgered by angry citizens who disagreed with forest plans or spraying herbicides. Ultimately, the citizens effected a profound change in the industry.

Public participation provided special-interest groups with a method for influencing government decisions. They quickly developed influencing into an art and science. Astutely employing participation techniques, they forced a change in the decision making of the United States Forest Service and forcefully inserted their opinions into the management of the national forests throughout the nation, particularly in the huge public holdings in the western United States. It is rumored that the citizen participation in

RARE II (Roadless Areas Resource Evaluation) in the late 1970s cost the Forest Service over $6.6 million to handle public input. Tens of thousands of comments from citizens poured into the Forest Service, which were analyzed by an elaborate coding scheme developed to count how many sentences, paragraphs or statements fell into each pre-defined category of comment.

The forest industry trade journals, particularly those dedicated to the primary processors that transform timber into lumber and panel products, generally express scorn for "activists" in retaliation for the scorn by activists complaining about forest industry practices. Many activists are in non-governmental organizations (NGOs), self-authorized, but with great influence domestically and globally. Famous movie stars or other luminaries lend the luster of their names to their causes. The full-page advertisement for an organization called "Mothers and Others for a Livable Planet," featuring movie star Meryl Streep as the founder, typifies the sophistication of their techniques.

Though activists were viewed as disturbing militants in the 1970s, by the 1980s they acquired social status. The 1977 *New World Book Dictionary* defined activism as furthering an interest "by every available means, including violence or warfare." By the 1986 edition activism had gained respect as "vigorous action in support of or opposition to a controversial issue."

Advocates in NGOs who speak in the name of the public exert more influence on public and private decision-making than their numbers suggest. The nuclear energy industry, biotechnology developments, and cigarette smoking each have their anti-establishment activist organizations. "Anti-establishment," as used by the federal government in describing forms of citizen participation, denotes organizations attempting to alter political power patterns and allocation of resources.

If we recall television programs like *Ozzie and Harriet* of the 1950s and '60s, we notice that the majority of middle-class Americans subscribed to a comparatively narrow range of values. Today, the American social fabric consists of thousands of niches, each with followers who share a few common denominators in beliefs and behaviors. These niches have inserted themselves into the American scene as new stakeholders, sharing an intense concern about the environment, government, industry, education, and business.

STOCKHOLDER INFLUENCE DWARFED BY STAKEHOLDERS

Stockholders finance the expensive equipment and operations of the forest industry, but stakeholders wield significant influence. Stakeholders include far more than the landowners, lumber buyers, lumberyards, builders, and manufacturers who depend on wood, their employees, and employee unions. The stakeholders outside the industry include practically every citizen who believes that he or she is entitled to clean air and clean water provided by forests. Nature lovers who love the forest for its own sake, biologists, ichthyologists, and ornithologists, all share a stake in the forest. Hunters, hikers, skiers, campers, and snowmobilers also believe they are entitled to having a say in the use of forests.

Farmers, communities, and homeowners in and around the forest, road builders, dam builders, and park builders all want their share. Citizens and governments of all types claim an interest.

The stakeholders who become advocates for a cause are a powerful force. Anti-nuclear activists, whose influence the nuclear industry at first failed to recognize, fought and nearly decimated the nuclear industry in the United States. Activists who believe we shouldn't fool with Mother Nature are questioning some new advances in genetic engineering. Based on the fear that genetic engineering may create altered microbes that could disturb the environment, Jeremy Rifkin of the Foundation on Economic Trends — a longtime opponent of genetic engineering — filed suit in September 1983 that provoked government officials to look at the loopholes in the guidelines for genetic alteration. The new stakeholders have a profound influence on the forest industry. Some activists interested in saving forests acquire minimum shares of stocks in the industry, gaining the privileges of stockholders.

THE DISCONNECTED FOREST INDUSTRY

The "forest industry" is an imposing collection of businesses that depend on forests. Forest managers and loggers; sawmills and plywood mills who process logs to lumber or panels; manufacturers who produce wood products, furniture, cabinets, pallets, floors, windows; brokers, wholesalers, and retailers who sell lumber to consumers — do not understand their place in the forest industry world.

The more than one million men and women who log timber and produce lumber and wood products know little about the significance of the millions of wholesalers, brokers, and retailers who bring these products to the market. City woodworkers think lumber comes from a truck. Jon Zeltsman, of the New York City Industrial

Technology Assistance Corporation (ITAC), points out that, "As milk doesn't come from the store, neither do the moldings, cabinets, or furniture in your house." He identifies the lack of recognition of the value of the industry as a problem "not only on the part of the consumer, but also from our political leaders, economic development policy-makers, educators, and some of our forest resource people."

One hand of the forest industry often doesn't know what the other hand is doing. Loggers don't worry about the needs of ultimate consumers, while the retail end of the industry knows the needs but doesn't understand the loggers' problems. The result is that these disparate segments do not speak with one voice. The existing segments are interdependent, but the dozens of sub-industries and businesses comprising the forest industry operate independently, with their own trade associations, trade journals, and industry standards.

In the winter of 1996, about fifteen hundred people convened the historic Seventh American Forest Congress in Washington, D.C. The Congress discussed forestry and ecosystems, but not processing timber or making cabinets or pallets. Following the Forest Congress, in March the Annual Sawmill Technology and Clinic, a major industry event, met to tackle the problems of converting timber to lumber in sawmills and panel plants, but paid little attention to the forest base.

A week later the builders of the nation's homes and offices — major users of wood — met to consider alternative building materials, paying scant attention to the forests. At yet another convention, the country's leading architects mulled over design, leadership, and technology for building and ignored the forest base supporting one of their essential materials.

Loggers who turn timber into logs with their chainsaws rarely connect with the lumberyards who sell the lumber produced from these logs. Sawmills who convert those logs into lumber are isolated from the home centers that sell wood products. The neighborhood cabinetmaker down the street is as much a part of the forest industry as the manufacturer who builds ready-to-assemble cabinets for sale nationwide. Unfortunately, neither feels a kinship with the loggers who harvest the timber.

These separate units fall into two organizational patterns: large and small.

The first group includes a few large, diversified (often international) corporations, that answer to the interests of their stockholders. Many are vertically integrated from the forest to the retail sales of a broad assortment of wood and paper products and even non-wood building materials. As forest resources become scarce and as wood products compete for customers with other materials, these industry giants must diversify to survive.

The extract from the 1995 Annual Report of Willamette Industries, a medium-sized giant in the forest industry, reveals a diversity that would amaze the founders of this company when they installed their first sawmill ninety years ago. Willamette Industries owns 1.25 million acres of timberland, providing about 40 percent of its sawlog needs. Sixty percent of their timber needs are purchased from other landowners. Nearly 14,000 people work in their ninety-five plants and mills throughout the United States. In 1995 alone they spent almost $500 million on projects to produce highly diversified products, including computer, copy, and printing papers, kraft liner, corrugating medium, bag paper, fine paper, specialty printing papers, corrugated containers, business forms, paper bags, lumber, plywood, particle board, fiberboard, laminated beams, and engineered wood. They use recycled urban wood and recycled boxes. Tree bark and wood particles not used provide 60 percent of the energy used in the plants.

The American Forest and Paper Association, whose 425 lumber and paper company members produce 95 percent of the paper and 65 percent of the United States' solid wood production, is a major association of large companies that play an important role in the forest industry, especially in manufacturing the wood-based products that require millions and millions of investment dollars.

The majority of small entrepreneur owned and operated companies rely on the availability of forest resources for their continued existence. These lumber mills, cabinet and furniture makers, pallet manufacturers, and other wood product producers are equipped to process wood and cannot easily switch to other raw materials. Since they are usually located in forest-dependent rural areas, these companies are often the prime economic and employment pillars of their communities.

Thus, the large, "greedy" corporation in the forest industry is a myth that dies hard. The 1992 United States Census of Manufacturers shows that 80 percent of the forest industry consists of small businesses with less than twenty employees. According to data compiled by the USDA Forest Service in September 1993 (Forest Resources of the U.S., GTR-RM-234), ownership of the forests in the United States is largely in the hands of small Non-Industrial Private

Forest Owners (NIPFs). The entire forest industry itself owns only 14 percent of America's forests. The large forest industry producers buy their timber from many of these small NIPFs in addition to their own. These small woodlands are gaining increasing significance as timber suppliers as the national forests become off-limits to harvesting. The ownership of forest land in the United States is shown in this table:

Forest industry	14%
Other private forests	59%
National forests	17%
State, Indian, other public	10%

The same Forest Service report on Forest Resources of the U.S. shows that lands owned by the forest industry are intensively managed. Although they comprise only 14 percent of the forests, they produce one-third of the timber harvested in the United States. The small private owners hold almost 60 percent of the forests and produce almost 50 percent of the total harvest.

For detailed information on forest ownership in the United States, the concise volume *American Forests, 1996,* by Douglas W. MacCleery, assistant director of the Timber Management Staff of the U.S. Forest Service, obtainable from the U.S. Forest Service, is invaluable. Of special interest to this *Reinvention of the Forest Industry* is the history of how the deteriorating forest and wildlife situation at the end of the nineteenth century led to the first environmental movement.

The forest industry culture represents the diversity of nature. The Northeast, Southeast, Midwest, Mountain States, and Pacific Northwest are home to tree species that thrive in the climate, seasons, soil, and other variables peculiar to each region. For example, a Pacific Northwest sawmill fed with larger conifer logs produces more lumber a year than an Eastern sawmill relying on smaller hardwoods.

The industry also includes support services that derive their income from forests: colleges of forestry, the United States Forest Service, the Bureau of Land Management, state foresters, forest research scientists, engineers, and equipment manufacturers. Major segments of the industry have their own professional societies and associations: the Society of American Foresters, the Forest Products Society, the Society of Wood Science and Technology, the American Pulp and Paper Institute, and trade associations. These associations, at least one for each wood product, are all dependent on the forest resource.

> DES PLAINES, ILL. — I like watching freight trains. This morning, though, I'm annoyed because the gates are down, and now I will be late for work. Stuck at the crossing, I begin to think about what is on this train.... And then I realize I like watching freight trains, especially those laden with building materials.
>
> The train invades my senses.... I can smell its cargo. A light rain has dampened the lumber, releasing the green, woody smell of the studs and plywood on the flatbed trailers.
>
> ...Seeing the freight makes me wonder about the people who loaded that lumber on the freight train. I think about the loggers who felled the trees and the mill that turned the trees into usable building products....
>
> ...I think about the forklift operator who will off-load the material and the truck driver who will deliver part of this load to a homesite from a lumberyard. I think about the builder who bought the materials and will construct a house from it. I think about all the people who had a hand

in creating this pile of studs and I wonder if they ever
think about their impact on this nation's economy and
their small but important role in creating shelter. Will the
Northwest logger ever meet the Midwest home buyer?
(Editorial by James D. Carper, *Professional Builder*, June
1996)

The forest industry reaches into the lives of every individual and every community, but most Americans'
knowledge of forests is composed of three-quarters myth
and one-quarter supposition. Recent surveys of college
forestry students across the United States revealed a frightening ignorance of the industry. The allegations about the
United States running out of forests have been so persuasive
that the majority of forestry students believed that 40
percent of current forests will be lost by the middle of the
next century! Likewise, they were convinced we were
depleting trees to make paper. The majority thought that
non-wood building materials, such as brick, concrete,
aluminum, plastic, and steel, have less impact on the environment than wood.

These assumptions fly in the face of the facts: The
records show that standing timber volume in U.S. forests is
actually increasing. The Forest Resources of the United
States' 1993 report illustrates dramatically that the standing
timber in the United States is actually increasing. Counting
all public and private owners, the timber volume per acre
has increased by 22 percent since 1952. It has increased
dramatically in all regions except the Pacific Northwest —
about double in the Northeast, almost double in the South,
about 20 percent in the Rocky Mountains. Despite the
heavy harvesting in the Northwest, the Pacific Coast
standing timber had decreased only 6 percent by 1992.

Probably more surprising is the relationship of the timber growth to harvesting. MacCleery writes: "In 1920 timber harvest rates in the nation were double the rate of forest growth; ... By 1986 net annual growth was 3.8 times what it was in 1920. In 1992 net growth exceeded harvest by 34 percent."

The American Pulp and Paper Institute reports that the volume of recycled paper used in U.S. paper production now exceeds 50 percent.

Impact of any material on the environment is measured by comparing Lifecycle Analyses. In recent years, European and American scientists have undertaken serious studies in Lifecycle Analysis, sometimes called the "cradle-to-grave" analysis of the environmental effects of building products before and after they become part of a building. An extensive Environmental Resource Guide, compiled by the American Institute of Architects, contains data on the energy required for the production and use of cement, plastic, brick, lumber and other building materials. The analysis considers the environmental impacts of manufacturing the materials, installation and use in buildings, eventual reuse, recycling, or disposal. The energy consumption (measured as embodied energy) in each step of the lifecycle is one of the most significant measures of environmental impact.

Comparing all the building materials, the massive study concludes: "Wood is a renewable resource, but proper forest management and timber harvesting practices must be employed to ensure adequate supply for future demand.... The embodied energy of wood framing is reported to be 91,618 BTUs per cubic foot.... The embodied energy of wood is substantially less than the embodied energy of potential substitute building materials." ("Wood Framing Lifecycle Analysis," page 1, *Environmental Resource Guide,* 1996, American Institute of Architects.)

All building materials originate from materials extracted from the earth by mining, drilling, digging, or harvesting. There is a finite supply on earth of most building materials. Trees and crops are the only renewable resources.

Each sector has its own trade association, who meet independently and separately; its own trade journals, terminology, and professional groups. Local cabinetmakers, for example, usually buy their lumber from a truck or a retail lumberyard and have little relationship to the loggers and sawmills who process the logs to lumber. There is no national meeting in which the growers, processors, distributors, and users of wood to manufacture products get together to discuss industry matters.

Using U.S. Census data on the number of establishments in each of the sectors, and based on the content of trade journals and trade meetings experience, it appears that fewer than one-quarter of the workers in the forest industry understand its significance to the United States' quality of life or the economy.

ENVIRONMENTAL PERFORMANCE —
THE NEW LITMUS TEST

The industry has awakened to the dismaying realization that technological miracles and good products do not earn public forgiveness for perceived or real environmental offenses. Chemical companies are not absolved from polluting because the chemicals they produce are essential for growing food and maintaining health. Activists will not condone producing wood for shelter as justification for clearcutting and other forest industry practices they dislike. Animal rights advocates will not sanction killing animals to produce coats just to keep humans warm. One consequence of a politically involved society is that technological advances are not accepted in lieu of social responsibility or a social ethic.

With this new trend in mind, environmental performance is a powerful new litmus test for the acceptability of companies and their products. Portraying their businesses as "green" has become a bottom-line factor for many companies, with major implications for the image of entire industries.

The Vision Statement of Quad/Graphics Inc. illustrates this new business practice:

"Our goal is wise, balanced use of all resources, including financial ones. We conserve raw materials, and we continually minimize waste and reduce our effect on the environment. Giving loving care to Mother Earth is good business."

DEMANDS FOR MORE BUSINESS SOCIAL RESPONSIBILITY

Public opinion no longer regards the environment as isolated or separate from a company's business practices; environmental performance directly links a company's social responsibility and the well-being of society to the bottom line.

At a meeting of the World Business Council for Sustainable Development held in Zurich, Switzerland, a Swiss industrialist, Stephan Schmidheiny, posed a pointed question: "Are the [world] financial markets," he asked, "opposed to this thing called 'sustainable development'? That is, are they opposed to economic progress that meets today's needs without making it impossible for future generations to meet their own needs?"

A task force assembled to answer this question found that the world's financial markets are quietly shifting to including environmental values. More than seventy-five banks signed a "Statement by Banks on the Environment and Sustainable Development." The first principle of the statement maintains: "We believe that all countries should work toward common environmental goals."

Actions previously considered "trendy" are now mandated as business social responsibility. Joel Makower and the organization called Business for Social Responsibility reports in *Beyond the Bottom Line:*

> It wasn't long ago that a company could issue an environmental mission statement, initiate a modest paper recycling program, perhaps ban polystyrene coffee cups from its premises, and stand a reasonable chance of making the nightly news or garnering a few column inches in the *Wall Street Journal*. Now companies that aren't doing these things are much more likely to find themselves fielding calls from reporters and activists....

Heightened environmental awareness and opportunities for stakeholder action are forcing all organizations, including businesses, into social and environmental responsibility. The magazine *Business Ethics,* now in its tenth year, states as its mission:

> ... to promote ethical business practices, to serve that growing community of professionals striving to live and

work in responsible ways, and to create a financially
healthy company in the process.

As environmental awareness emerged in the last
quarter of the twentieth century, social responsibility
promises to be the cornerstone of the twenty-first century.
The new definition of any institution's role in society
includes assuming greater social responsibility in addition
to its traditional mission — the production of goods and
services to meet the needs and wants of customers.
Heretofore, customers weighed their needs and wants
against price and quality. Today, many consumers add environmental and social acceptability to the attributes they look
for in products.

Peter Drucker maintains, "Organizations must compe-
tently perform the one social function for the sake of which
they exist — business to produce the goods, schools to
teach, and hospitals to cure the sick.... They can do so only
if they simplemindedly concentrate on their special mission.
But there is also society's need for these organizations to
take social responsibility...."

Many companies have debated the wisdom of taking
the lead in solving the world's problems, despite Peter
Drucker's advice to focus exclusively on their own operations. Organizations and institutions no longer are free to
debate that decision; there's no room for any more debate.
Social and environmental responsibility have already
become engraved in the hearts of business and in their
customers who consider this social responsiveness an additional product attribute.

TWO PLUS TWO EQUALS TEN —
THE NEW ARITHMETIC

Passing the environmental litmus test and social responsibility test is like adding two plus two, only the answer isn't four; it's ten. However, the equation takes on so much complexity that it's difficult to pass both tests at once.

The forest industry, still struggling to manage the responsibility for the total environment which has been thrust upon it, now faces the added responsibility for providing workplace diversity, community involvement, employee empowerment and family welfare, saving the rainforests and endangered species, and other responsibilities which are still undefined.

A long-distance telephone marketing company promises big savings in telephone bills.

Two-Million-Dollar Giveaway
from Working Assets.

Every time you call Working Assets Long Distance, we give one percent of your charges to nonprofit groups working for peace, human rights, economic justice, and the environment. And if you don't consider one percent very much, consider this: In 1995 alone, we generated more than $2 million for nonprofit groups selected by our customers. Groups like Greenpeace, Human Rights Watch, Planned Parenthood, and Children's Defense Fund. This year, with the addition of new customers like you, we hope to give away even more. (Advertisement, Working Assets)

"But," they say, "we do more than that. We also save rainforests, endangered species, and even lives." They add,

"While saving money is nice, we believe helping save the world is the only 'savings' that truly qualifies as 'big.'"

The question now is "How?" rather than "Why?" With the new arithmetic, social demands are not viewed as irrelevant distractions to a business's major purpose. Society no longer is willing to bear the "social costs" of breathing polluted air, or soil contamination, or dwindling fish reserves due to water pollution from an industry. Instead, society wants to shift these costs back to the firms that caused them — making the polluter pay.

A company that caters to consumers is especially vulnerable. Ben and Jerry's, the environmental ice cream company, touted as a model of modern social responsibility, is criticized in *Inc.* magazine by another socially responsible ice cream producer, Mayas Ice Cream, for not carrying that responsibility deeper into the company's operation. "Corporate social responsibility sounds easy when you're saving the rainforest," the author says. "It gets more complicated when you're coming up with a maternity leave policy or deciding about a dress code."

Advocates who play leading roles in making society's rules make their voices heard in the boardrooms and the washrooms. Speaking in the name of the public, they insist on receiving more value from institutions than the products or services they deliver. A Midwestern sawmill operator's comment represents the frustration of many of the smaller operations in the forest industry: "All that social stuff costs something — and then they complain about the price of lumber!"

Executives worry that all this social responsibility ignores the bottom line and makes stockholders unhappy. "I am socially responsible because I make a profit," a business owner protested. "I have to be profitable in order to provide jobs at fair wages and contribute to the economy of my community. What more do they want?"

Like Simmons running his sawmill, many businesses have justified their existence by simply providing jobs and helping the community. Some have been forced out of business because their practices were not acceptable: nuclear plants closed by fear; chemical factories closed to reduce pollution; Western sawmills — the backbone of rural communities in five Western states — closed to preserve spotted owls; defense plants and papermills have all been sacrificed for the sake of an ethic considered more important than the products or jobs they produce. In the early 1990s the public was not impressed that more than 15,000 loggers, sawmill and plywood workers' jobs in Oregon were sacrificed on the altar of social responsibility.

Executives like Simmons want to be socially responsible, whatever that means, but they object to outside pressures interfering with their internal decisions. One solution is for businesses to attempt to create an image of responsibility by using sophisticated advertisements. Programs like planting a tree in a customer's honor aim to develop the image of environmental responsibility. A reputation for social responsibility helps attract good employees and stockholders. The *Job Seeker's Guide to Socially Responsible Companies* lists 1,000 companies that pass the social responsibility test. Resource books for investing in socially responsive companies help socially aware investors put their money where their heart is.

APPEASING THE PUBLIC APPETITE

The road to social responsibility is not clearly marked because the rules evolve slowly. Most consumers still buy because they need or want a product and the price and quality are right, but surveys repeatedly show that about 15 to 20 percent of the population buy with their conscience as well as their head. In other words, there are educated consumers who earn and spend goodly sums of money and care about the environment and society. Their influence is being felt throughout the business world.

An industry that incurs the displeasure of the public is likely to feel it on the bottom line. Reactions occur from many arenas: magazine articles, television shows, advertisements, movies, actions by celebrities, books, school texts and other communications deliver criticism that eventually hits the institution in its pocketbook. Consumer boycotts of products may directly decrease sales. There are campaigns to develop subtle, subconscious feelings of guilt that the customer will harm the environment if he buys certain products. The consumer may be made to feel guilty of causing the chainsawing of trees if he uses paper towels, or paper bags, or buys a Sunday newspaper.

"Guilt by buying" has made its mark on the forest industry: The campaign to save trees by not using wood or paper has been effective.

The author of a story for the *Southface Journal,* published by the Southface Energy Institute, demonstrates the effectiveness of guilt by buying when purchasing lumber to build his new home:

> For a long time, this moment, which should be very gratifying, has filled me with guilt. Being concerned about

the environment, all I can see is a huge mass of lumber —
11,000 board feet in the average home.... When you
consider all the homes being "stick-built" in this country,
the amount of dimensional lumber used is staggering.
More than one-third of the lumber used in the U.S. goes
into building new homes. It's time we took a long look at
the way we approach home building.

THE FOREST INDUSTRY ON THE PRECIPICE

The forest industry is standing on the precipice, ready
to dive into a major redefinition of its role in society. Many
of the largest forest industry companies have already
decided to plunge into the environmental correctness sea.
When the American Forest and Paper Association adopted
its Sustainable Forestry Initiative in 1995, it marked a turnaround as significant as the environmental "Earth Day."

After years of nagging by environmental advocates,
the forest industry — in a turnaround that has some industry
leaders frothing and others cheering — adopted sustainable
forestry principles which reflect environmental values.

The principles of the Forest Stewardship Council, a
leading international environmental organization; and those
adopted by the American Forest and Paper Association, a
leading association of American forest industry companies,
include similar commitments.

The American Forest and Paper Association
Sustainable Forestry Initiative supports: "The practice of
sustainable forestry [to integrate practices that are environmentally appropriate], a commitment to society, and a vital
forest-based economy."

The Forest Stewardship Council declares that its
mission is: "to support environmentally appropriate,

socially beneficial, and economically viable management of the world's forests."

The American Forest and Paper Association Board of Directors approved their landmark principles and guidelines in October 1994; the Forest Stewardship Council approved its basic document in September 1994. The two organizations share similar goals, objectives and principles. While each organization has similar goals, the differences focus on how to achieve these goals.

Timber Group Sets Environmental Criteria

WASHINGTON — Major U.S. wood products companies spent hundreds of millions of dollars last year putting into place environmental standards aimed at protecting timberlands, an industry trade group said Thursday.

The American Forest and Paper Association said participation in its "sustainable forest initiative" is now mandatory for membership, and twenty-seven companies have quit the group or have been suspended for failure to comply. "That is the price of putting teeth in our program," said W. Henson Moore, president of the trade group.

Although the short-term costs are high, industry executives said, companies needed to do something to assure the public they're determined to protect the environment.

Some environmentalists praised the industry plan as a step in the right direction.... Other environmentalists had problems with the proposal and said it conflicts with industry's position for continued logging in old-growth forests in the Northwest. (*Bloomberg Business News,* April 12, 1996)

Reflecting changes in attitudes and legislation, the forest industry and the environmental community have become more pragmatic, accommodating Americans' concerns for the environment and recognizing the necessity for a robust economy.

The public's distrust of the forest industry's sincerity as environmental stewards of the forest remains a palpable barrier to solving the problems facing the world. Furthermore, the forest industry's tendency to derisively label activists as "enviros" and eco-extremists fortifies the barrier. In these changing times, the forest industry knows its survival is at stake; environmental advocates believe the survival of the earth is at stake.

The solution isn't easy. The public interest demands that both sides set their sights on their common goals and forgive each other's real or perceived past trespasses. The pressure is on the forest industry to reinvent itself to meet the public's environmental and social expectations. On the other side of the debate, the environmental community has a responsibility to determine how to accommodate both the environment and people.

Companies, environmental advocates and research groups are discovering that working together, rather than at cross-purposes, produces results. An association of businesses based in New York City called the Conference Board reported that alliances of otherwise disparate groups are now being formed rapidly and are producing win-win results when a problem is approached with a desire to work together. In the late 1980s, the 175 companies in the non-profit Chemical Manufacturers Association (CMA) — many who usually compete with each other — decided to take a significant step toward satisfying the public's desire for both useful products and a safe and clean environment. CMA members account for 90 percent of basic chemical

_.._

productive capacity in the United States and Canada. Their decision resulted in the creation of the "Responsible Care" program — an unprecedented industry commitment to continuous improvement in environmental performance to gain the trust of the public in chemical manufacturers. The program is being vigorously pursued and is credited with making marked improvement in industry performance.

The forest industry is on the brink of a new role in society, ready to demonstrate that it can be trusted to perform in an environmentally and socially appropriate manner.

Chapter 2

Paradox in Paradise

Evelyn and Allen Ridell loved the place, an old house, cupped in the gentle Pennsylvania hillside, at first sight.

"Just the size we want," Evelyn exulted. "There's room enough for the whole family to visit."

"But, come, look here," Allen pulled her around to the back of the house.

In unison, they let out a contented, deep sigh, standing silently, arm in arm, their eyes hungrily taking in the thick forest sloping from the house. Leaves blazed orange and red and yellow from the first touch of frost, the trees magnificent in their fall glory. Evelyn clasped her hands as if to hold the moment forever.

"It's as close to paradise as we'll ever get. Untouched. Imagine having breakfast every morning looking at this view!"

Allen, forever practical, said, "I wonder if the breakfast room faces this side."

"I don't care; we'll change it if we have to. But I want this house and this mountain and those trees. Think what it will be like in the spring!"

Evelyn and Allen bought the mountain and the trees and the fawn who casually wandered in the yard and, incidentally, the house.

Though the house turned out to be in not as good condition as they hoped, they settled in happily, watching their forest change from bursts of color to snowy stillness, to the explosion of spring, and the bustle of summer.

IN THE BEGINNING

On a brilliant fall morning five years later, as they watched their fawns playing hide and seek in the trees, a strange animal rumbled through their forest — a huge machine tended by a crew of men wearing hard hats. Alarmed, the Ridells climbed down the fragile path they had created through their forest. "What are you doing to my forest?" Allen shouted to the crew boss.

"It's time to cut it, according to the schedule," the man who appeared to be foreman replied.

"You can't cut my forest," Allen shouted back. "It's a sin to chop down even one tree. Some of these trees have been here since before the Pilgrims landed."

Noting the tears welling in Evelyn's eyes, the foreman answered quietly.

"I don't know where you folks got your information. None of these trees have been here even more than fifty or so years. This is cutover land where the forest has regrown — over and over. Anyway, this isn't your forest. This land belongs to Henry Jones. Been in his family since his great-grandpa settled here. I suggest you folks check your deed."

Allen answered angrily, "I don't care who owns the land. This is our forest. They're our trees. It's our view. What right have you to cut down what God provided us?"

The foreman, deciding not to debate the issue, replied, "Well, nice meeting you. We have to get back to work. Jones needs the money now to pay his taxes."

Evelyn was bewildered. "How did we get into this mess?" she asked Allen.

Pulling at Evelyn's arm, he turned to climb back to the house. "Hurry up. We'll get an injunction to stop the logging while we straighten this out."

Allen and Evelyn's experience mirrors the paradox in the forest paradise. When a forest is logged, paradise looks more like Armageddon. Incidents similar to Allen and Evelyn's occur all over the United States. Often former city dwellers anxious to get close to nature discover that their forest is someone else's property. They buy for the view of the forest, and then some logger destroys the view. The conflicting objectives of saving trees to enjoy for their beauty and using (logging) trees to provide needed creature comforts are at the root of the confrontations that are crying for solution.

Simmons, the sawmill owner, had once asked, "How did we get into this mess?" when informed that he couldn't log his national forest in Idaho. It was public land. Across the country, in their rural Pennsylvania haven, Evelyn asked "How did we get into this mess?" when the logger's chainsaw entered their forest.

The "mess" is the conflict between uses of the forest — timber or scenery, fresh air and water, hunting or camping, or houses and furniture, wildlife habitat or paper.

For the Ridells and their friends, the ideal forest is "natural"; that is, untouched by humanity. Because humans need what the forests provide and there are considerably more people since Adam and Eve followed the dictate to have dominion over the forests, natural forests are scarce in the world. Like the Ridells, few people can distinguish a natural primeval forest from a second or third growth.

Since the earliest days of recorded history people have believed that resources were put on the earth for man to use. Permission and authority to cut trees stems from the Old Testament of the *Bible*. In Genesis God instructed man to be fruitful and increase in number, to fill the earth and have dominion over it (Genesis, 1:28, 29). Rudyard Kipling's stirring *Recessional* poem boasts about "dominion over palm and pine."

However, even a Biblical precedent can backfire. The apparently wasteful and wanton destruction of natural resources to fill the needs of a prospering nation turned many Americans away from belief in their right to have dominion over the earth.

In the 1960s, man's dominion over the earth was challenged by a new version, the environmental-ecological paradigm or model. Humans, according to this view, are simply part of the earth's ecosystem. One faction supporting this ethic maintains that human needs or rights are not superior to but comparable to those of trees, owls, wetlands or snail darters. According to this paradigm, civilization is an intruder in the natural order, a despoiler of the ecological balance of the earth.

Problems arose when unrestrained dominion over the earth became an unsustainable folly, not in harmony with nature. Chris Maser, scientist, philosopher, forester expresses the essence of this view: "The enlightened people of Western civilization have forgotten our place in the Universe. Whereas primitive people knew who they were and where they belonged in and of creation."

The history of civilization is the story of the conquest of the natural world. Until this century, the natural world appeared to be unlimited. We came, used what we wanted, abandoned what we didn't want — often in a messy condition — and then left.

The conflict between the new and old beliefs is about control of the world's resources. The dominion model or paradigm puts man in control. The environmental paradigm places man on an equal footing with other members of the ecological system.

This dramatic shift in the views of man's relation to the earth and nature has defined the last third of the twentieth century even more than technological advances. It has had a greater impact on the forest industry than natural disasters or unnatural events.

CHALLENGING AN ANCIENT PARADIGM

Evelyn and Allen concerned themselves little with the philosophical arguments. They just didn't want their forest cut. They shared the religious spiritual emotion about forests and trees expressed so eloquently by Henry Thoreau and other philosophers.

Outraged at the "desecration" of their forest, they decided to join a local chapter of the Sierra Club, who helped them organize the Homeowners Forest Alliance, covering the entire perimeter of their forest, "to nip any further logging in the bud." They enthusiastically embraced environmental advocacy. Evelyn Ridell complained, "We're so busy attending meetings and planning strategies with the Alliance that we hardly ever have time anymore for those leisurely breakfasts watching the deer play."

The new paradigm changing man's place in the universe plunged the forest industry into a dilemma that has evaded solution to this day.

The phenomenon that has come to be known as the *environmental movement* grew out of this new view. It is a movement that has far-reaching impact, markedly affecting all the industries that extract natural resources from the earth. It has changed the course of the industrialized countries with major manufacturing facilities, largely dependent on mining, extracting, or harvesting natural resources from the earth to convert to products people need.

Much of the international environmental movement's current focus is to bring environmental and social equity to the underdeveloped non-industrial world.

Everything that happens to the forest affects every segment of the forest industry and everyone on the planet. The environmental ethic that engulfed the consciousness of the country changed the way the forest industry conducted its business and stirred a pot of conflicts between the public and the forest industry.

In the immediate post-World War II years, the nation depended heavily on the forest industry for lumber for the expansive economy. Expanding population, expanding prosperity, and expanding influence in the world fueled demands for more amenities, more freedom of action, and unending progress. Babies were a welcome sign that all was well: A woman pregnant with her fourth child was congratulated for contributing to our continued economic prosperity.

"Progress is our most important product" boomed authoritatively in the new television commercials, reflecting the buoyancy of the period. Success was defined as acquiring more — a second bathroom, a second car, a second home, a larger office. Acquisition was the basis of a welcome prosperity after the deprivation of the Great Depression and the stringency of the World War years. Having the money to acquire and the goods available to buy proved that the lean years were over. School children were taught that science could conquer present and future problems, and their parents were also happy to embrace this hopeful outlook.

But not all scientists were satisfied with this acquisitive behavior. In 1948, on the heels of World War II, a world-famous scientist named Fairfield Osborn warned in his book, *Our Plundered Planet,* of a worldwide conflict with nature with potential for "ultimate disaster" greater even than the misuse of atomic power. He wrote that our own "brutal, wanton destruction of earth's natural resources" was the greatest threat ever to confront mankind. Most Americans ignored the dire warnings of *Our Plundered Planet* in the optimistic post-World War II world.

FROM THE SPACE AGE
TO THE ENVIRONMENTAL ERA

In the 1950s, technology had invented the marvels of television, a device that would ultimately change our daily routines. Atomic power was hailed as the beginning of a new era, with unlimited energy for astounding new possibilities. The first atomic-powered submarine went to sea. The science-fiction reality of the first space satellite, *Sputnik,* and the conquest of space reinforced the conviction that technology was the universal problem solver. In 1950, the federal government created the National Science Foundation, furthering a belief that technocracy and cybernetics would solve all problems. Another innovation was the new jet passenger service that would shrink the earth's boundaries.

However, such unfettered progress was creating a sense of unease in some scientists and academics. "What will happen to this fragile planet," they asked, "if the current practices of using natural resources continue unabated?" Their writings warned of unspeakable horrors as they tried to convince Americans to consider the impact of progress on the environment. These future-oriented scientists focused first on curbing pollution. Smokestacks, heretofore considered a welcome signal that factories were operating and paychecks would be forthcoming, came into disrepute. Companies with logos that proudly showed curls of smoke emanating from their factories quickly removed the telltale plumes. The early Clean Air acts were responsible for the elimination of the teepee burners. The motto of the time became "Dilution is the solution to pollution," and taller smokestacks were built to detour the pollution into the heavens.

The suspicion that all might not be well in this ebullient new technology-centered era hit the world in 1956, with the publication of Rachel Carson's pivotal book, *Silent Spring*. Many scientists rejected her claims, but the fear that technology could also be an enemy crept into the nation's euphoria.

Citizens learned that individuals could effect change, and a generation of college students sharpened their activist skills by marching against discrimination as the civil rights movement fought racial discrimination.

The framework for thinking about nature in the years before World War II was illustrated by the persuasive movies of the Depression era, typically expressed by James Stewart in the classic *Mr. Smith Goes to Washington,* in which he urged the conquest of nature. However, by the 1960s, the still-simmering environmental paradigm began to poke into public discussion. The environmental paradigm proposed a new pattern of behavior. The old "dominion" version underpinning the rules by which post-World War II parents lived was fraying at the edges, yielding to a young counter-culture. Situational ethics replaced the dogma of the Biblical Ten Commandments, and school prayers were declared unconstitutional. The assassinations of John Kennedy, Martin Luther King Jr., and Robert F. Kennedy added to the growing instability. The unwavering patriotism of "my country, right or wrong, but my country," which bolstered generations in the first half of the century, gave way to distrust of the government and its institutions. Changes in the rules guiding love, courtship, family, and community alienated many youths from their parental generation.

Civil rights marches, Vietnam War protests, student takeovers at university campuses, violence against anti-war demonstrators, Moratorium Day, and women savoring new freedoms and new jobs jolted the complacency of the old patterns. The country seemed to be boiling with new and, often, young energy.

While the youth were questioning the government's authority, the federal government grew in power and numbers. Americans began to look to their government for redress of an agenda of wrongs and for more help in meeting life's problems. Out of this mood, Medicare was born in 1966.

Ordinary citizens, provided with access to government decisions by laws that ordered agencies to invite public participation in decisions, learned that activist alliances had clout. The country split into hundreds of groups, each pressing lawmakers to act on their special cause. Citizens discovered that the pen was still mightier than the sword, and they sharpened their skills at creating public opinion. Planning, a professional vocation for most of the century, became both the vogue and the law. Citizen participation transformed into an end in itself, with the process often looming as more important than the outcome.

Thus, the dual influences of the government's immersion in the everyday affairs of Americans and the increased involvement of citizens in federal government decisions acted as the midwife for the birth of the environmental movement. Without the willing acceptance of more government dictates by industry and by ordinary citizens, the environmental movement might have been an alligator with no teeth. Without the prodding of the environmental advocates and citizens participating in the government, there might not have been the impetus for increased regulatory activity.

FORMATION OF THE ENVIRONMENTAL DYNASTY

Starting first as a zephyr, the environmental paradigm grew to a hurricane by the end of the 1960s. Inspired by Aldo Leopold's *A Sand Country Almanac,* a spate of books appeared on the scene passionately exhorting Americans to preserve the environment. New stars appeared in the environmental universe: Robert Rienow, political science professor at the State University of New York, and his author wife Leona, penned *Moment in the Sun,* unveiling the health dangers of smog. Barry Commoner, director of the Center for the Biology of Natural Systems at Queens College, City University of New York, 1989, wrote *The Closing Circle* in 1972. Dr. William Shurcliff was a senior research associate of the Cambridge Electron Accelerator of Harvard University. In March 1967 he founded the Citizens League Against the Sonic Boom (CLASB) and became its director. The CLASB became a model for committees organized to oppose the sonic boom. Shurcliff wrote *The SST and the Sonic Boom Handbook* in 1970 and was credited with halting the building of the supersonic transport planes. Paul R. Ehrlich, professor of Biology at Stanford University, wrote the best-selling *Population Bomb* (1968) and numerous books on population. Kenneth Boulding, professor with the Institute for Behavioral Sciences at the University of Colorado, authored *The Meaning of the Twentieth Century.* Their books, frequently in paperback and relatively inexpensive, appeared in airport book displays and university bookstores.

As the movement gained momentum, Paul Ehrlich's powerful *The Population Bomb* stirred intellects. The Sierra Club produced its *Wilderness Handbook,* Friends of the Earth was founded, and Ralph Nader urged schoolchildren to action against the "deadly menace" of air and water pollution. In 1969, *America the Raped* was published, raking over the coals "the engineering mentality" in league with governments, scientists, industrialists, and the forest industry who "despoiled the world's most magnificent continent."

In tandem with concern about the environment and the willingness of government to enter into the business of solving the problems of its citizens, another movement pushed its way into the scene. This movement might be labeled Empowerment, which urged ordinary people to act and save the world from injustices and impending environmental catastrophes. Several private foundations supported a citizen involvement training project to teach citizens how to organize to win a rent strike, or stop a nuclear power plant, or right whatever wrong they felt confronted them.

Saul Alinsky's *Rules for Radicals* became chic reading for ambitious activists, who were often graduates of publicly funded organizations empowering activists "to stop a nuclear power plant or force rich people to pay taxes."

Although the Ridells, our Pennsylvania couple, who had earned their money as professionals, did not propose to force rich people to pay more taxes, they adopted the bold suggestions in *Rules for Radicals* in their Alliance and felt empowered to change the world.

The Woodstock festival captured the spirit of youthful rebels, to the dismay of many parents still supporting their children by working in the industries operating under the old paradigm. Charles Reich's *The Greening of America* urged a change in our basic premises, including "uncontrolled technology and the destruction of environment and the powerlessness of the current democracy."

The organizations devoting their activities to the environment grew to become a new dynasty. Since many of their actions focused on disciplining and scolding the forest industry for its perceived dissolute past, the industry came to regard them as the enemy. The forest industry holds the view that the environmentalists have seduced the media, lawmakers, and the public into joining their assault on the industry. This is the "mess," as Simmons and his colleagues view it.

Leaf through almost any forest industry trade journal, and an article on the environmentalists, or "enviros," catches the eye. To the forest industry, the environmental community has been the source of their woes for more than twenty-five years.

Of course, the Ridells and the environmental movement perceive the "mess" from a different perspective. Leaf through the magazines published by the Sierra Club, the National Wildlife Federation, the Izaak Walton League, the Audubon Society, or the publications of any of the Group of Ten (the ten leading environmental organizations), and the forest industry is painted as the "enemy."

The National Wildlife Federation publishes an annual *Conservation Directory* listing more than 2,500 conservation organizations, ranging from small local groups to the giant National Wildlife Federation. Some organizations boast an agenda of cooperation, education and research, typified by the Nature Conservancy. Others, like Earth First and Greenpeace, use civil disobedience to attract support. The environmental groups in between practice lobbying, litigation, confrontation, and demonstrations.

The powerful Group of Ten includes:
- National Wildlife Federation
- National Audubon Society
- Sierra Club
- Wilderness Society
- Izaak Walton League
- National Parks and Conservation Association
- Environmental Defense Fund
- Natural Resources Defense Council

The remaining two included in the Group of Ten vary with the interests of the list compilers.

Though people lump all these groups together as environmentalists, they differ in their missions and often scorn
each other's activities. The Izaak Walton League's publication, *Outdoor America,* carried an article about a "War in
the Woods," which covered a war declared by animal-rights
activists who taunt and jeer hunters in the woods. In 1993,
Vancouver, British Columbia, voters, convinced by a
passionate animal-rights campaign, decided by a 54 percent
majority to close the popular Stanley Park Zoo because wild
animals ought not to be displayed primarily for the entertainment of people. Stanley Park Zoo earned the distinction
of being the first major municipal zoo in North America to
be closed by the animal-rights movement.

An effective infrastructure known as NGOs, or non-
governmental organizations, supports the environmental
community. As grassroots organizations that emerged in the
1960s and 1970s, NGOs promote human rights, political
freedom, and environmental stewardship. More than 30,000
NGO representatives, about 17,000 of them from the United
States, participated in the 1992 landmark Rio de Janeiro
conference that catapulted the forest industry to the top of
the international environmental agenda. NGOs have
become a powerful force in propelling environmental
concerns in both underdeveloped and developed countries.
Ranging from local to international groups, NGOs now
possess the stature and influence to take their place
alongside government agencies in international delibera-
tions.

NGOs translate environmental considerations into
social concerns. They seek to achieve social equity or parity
between underdeveloped and industrialized nations,
between the "haves" and "have nots," and between the

North (representing industrialized nations) and the South, representing non-industrialized nations). They have acquired the power to shape the policies of government and intergovernmental organizations in linking environment and development.

NGOs affecting the forest industry come in three varieties:

Timber Trade Associations:
Business and Institutional Furniture Manufacturers Association, Hardwood Plywood Veneer Association, and Woodworkers Alliance for Rainforest Protection.

Environmental Organizations:
National Audubon Society, National Arbor Day Foundation, National Wildlife Federation, Sierra Club, and American Forestry Association.

Some environmental organizations have acquired reputations as radical activists — Friends of the Earth, Greenpeace, and Rainforest Action Network.

Professional Associations:
Society of American Foresters, American Institute of Architects, and Center for Environmental Study.

The 1995 U.S. Congressional dispute on budgeting government funds for activist or non-profit organizations was considered an attempt to withdraw monetary support for NGOs, much of it built in after creation of the Environmental Protection Agency (EPA) in 1969.

After passage of the National Environmental Policy Act (NEPA) in 1969, whose lofty purpose was a national policy "which encourages harmony between man and his environment, eliminates damage to the environment, stimulates the health and welfare of man, to enrich the understanding of the ecological system and natural resources important to the nation." The major impact of NEPA arose from the Section 102 (2)(c) requirement that all agencies of the federal government prepare detailed Environmental Impact Statements (EIS) on all major federal actions significantly affecting the quality of the human environment. In the words of an article in fall 1997 of *The Public Interest* on the Environmental Impact Statement versus the Real World, "...a major effect of the EIS requirement has been to give environmental groups a legal and political instrument to cancel, delay, or modify development projects that they oppose." By June 1975 there had been approximately 332 suits against federal agencies, and another 322 suits were still pending.

Many court decisions later, the EIS became a focal point in the issuance of licenses or permits to private companies for construction of projects that would affect the environment.

NGOs target industrial actions they believe harm the environment, or are not socially responsible. The chemical industry, forest industry, oil industry, and almost any other industry whose processes may affect the air we breathe, the water we drink, our health, and sometimes even our wealth, have received their share of NGO attention. In 1996, the Sierra Club voted two to one to oppose all timber harvesting on national forests, marking an important turning point for an environmental organization once considered moderate.

WASHINGTON — The Sierra Club for the first time is advocating an end to all commercial logging in national forests. Members of the century-old environmental group voted by a two-to-one ratio to change their timber policy.

...Members voted 39,147 to 20,287 in support of banning logging on all federally owned lands in the United States, a club spokesman said Monday.

...Regina Merritt at the Oregon Natural Resources Council said the addition of the Sierra Club to that list "demonstrates that people are increasingly concerned that we have lost our ability to protect ancient forests and salmon. I think it may be an indication that the salvage rider is backfiring." (The Associated Press, April 23, 1996)

THE SEEDS SPROUT IN FRIENDLY SOIL

The combination of a willing federal government, social unrest, and empowered citizens proved fertile soil for the seeds of the environmental prototype. First, the seeds flowered in the 1970s. Then the concept shifted in 1970, when the new environmental perspective of man and nature replaced the old dominion view.

In 1970, the publication of a new round of books was destined to translate the alarm of the earlier books on pending environmental disasters into action. The influential Sierra Club published *Ecotactics, The Sierra Club Handbook for Environmental Activists,* defining ecotactics as "the science of arranging and maneuvering all available forces in action against enemies of the earth." The enemies included DDT, progress, polluters, rich industry loggers, freeways, and government. Twenty-eight passionate ecologists, more than half under thirty years old, contributed their opinions to *Ecotactics.*

The same year, Friends of the Earth published *The Environmental Handbook,* replete with handy information for the first international environmental Teach-in held in April of 1970 to marshall public opinion about the urgency of the environmental crisis. The first Earth Day was a smashing success that firmly launched the new environmental era.

Ralph Nader contributed *The Vanishing Air,* a polemic against "the pervasive environmental violence of air pollutants [that have] imperiled health, safety, and property throughout the nation for many decades." Paul Ehrlich's earlier *The Population Bomb* inspired a fiction nightmare book, *Population Doomsday,* based on "the ecological perils which presently haunt the future of mankind."

In case anyone still had doubts about an environmental crisis in 1970, *The User's Guide to the Protection of the Environment* suggested saving the redwood forests by halting building with redwoods, using new materials made from industrial by-products, and building with pre-stressed concrete, metals, cinder blocks, and plastics.

The momentum of those heady years extended to the establishment of legal rights for natural objects. In the Mineral King Valley controversy, the U.S. Supreme Court defined being "aggrieved" to include aesthetic and recreational grievances as well as economic values. In this decision, trees were claimed to have standing before the courts.

Legislation affecting land, food, technology, forests, and significant social changes made their way through Congress. Forests, fish, water, air, streams, and wetlands were protected. Pollution from oil spills, radiation, noise, toxic substances, and other potential threats was forbidden. The number of such environmental acts enacted between 1960 and 1976 was just short of 100. The progression tells the tale:

1960 to 1964	18	3.6/year
1964 to 1969	27	5.4/year
1970	17	17/year
1971 to 1974	29	7/year
1975	4	4/year
1976	4	4/year

In 1970, landmark legislation was enacted with more environmental laws passed than in the years preceding or following. One could reasonably draw a line in time and say of 1970, "This was it. This was when the environment became a movement that changed America's culture." The turbulent student resistance to the Vietnam War, with events at Kent State and marches on the White House, absorbed the public, distracting attention from the environmental shift taking place.

The new Environmental Protection Agency (EPA), with the authority to administer twelve major environmental laws that touched on every aspect of the environment, occupied center stage in 1970. As a result, commented the chemical journal *Chemical and Engineering News* (October 30, 1995), on the twenty-fifth anniversary of the EPA, "...the actions of no other federal agency so directly affects the lives of so many Americans. We all breathe the air. We all drink the water. We are all exposed to the ubiquitous chemicals of modern society." In the scientific community there is general agreement that EPA has done its job well in attacking filthy air, contaminated water, and persistent pesticides.

The effective Natural Resources Defense Council, organized to "protect America's most endangered natural resources;" and the Environmental Policy Center, were two of many environmental-advocate organizations born during the era. The Center for Science in the Public Interest, destined to be a mover and shaker in protecting the environment, sprouted. Old-timers like the Sierra Club, the National Audubon Society, the Izaak Walton League, and the National Wildlife Federation flexed their muscles with renewed energy.

These views were supported in 1972 by the Club of Rome, a group of scientists and industrialists from many countries who were convinced that the breakdown of society and the irreversible disruption of life on this planet was imminent under the old dominion concept that man ruled the world. The Club of Rome wrote *A Blueprint for Survival*. The *Blueprint* predicted complete disaster by the end of this century. Declaring that our (then) current industrial way of life is not sustainable and that radical change is necessary and inevitable "because ... the present increases in human numbers and per capita consumption ... are undermining the very foundations of survival," the *Blueprint* convinced many intellectuals. The *Blueprint* proposed a strategy of seven major changes in the rules and regulations that governed the behavior of society; it redefined boundaries for survival; and it defined new frugal standards of behavior within those boundaries. Truly, a major new movement was born.

"Better things for better living through chemistry," proudly advertised in the 1950s and 1960s, was decried as unsustainable, according to *A Blueprint for Survival,* that will "impose hardship and cruelty among our children."

Although the *Blueprint* did not immediately change the old rules, it did inspire a critical mass of brainpower and effort to urge creation and adoption of the new rules. A torrent of literature poured forth from universities, think-tanks, and new organizations on the theme that growth for growth's sake is destructive and unsustainable.

Most of the *Blueprint* proposals for surviving the imminent catastrophe urged curbing consumption of material goods and adopting an alternative, more thrifty, mode of consumer behavior. The majority of Americans have not yet bought into the Spartan program, but they are nibbling at the edges of curbing consumption, as the rapid growth of recycling and reuse indicates. Although Americans are still reluctant to give up automobiles and television sets, the nation has largely adopted the idea of protecting the environment. The culture of America is now environmental.

Between 1970 and 1975, federal "economic" regulatory agencies increased by 25 percent, and the number of federal "social" regulatory agencies increased by 42 percent! By 1978, industries contributing to over 20 percent of the Gross National Product (GNP) were considered "heavily regulated": cement plants; nitric acid and other chemical producers; petroleum, lead, iron and steel mills; aluminum reduction; kraft pulp mills; and other industrial plants. The dramatic increases resulted in a burst of new professions — compliance specialists, public affairs scanners, lobbyists, lawyer specialists in specific fields, and corporate planners.

Through the 1970s, provocative ideas supporting the new movement poured from writers and leaders throughout the United States and the world. Herman E. Daly's *Toward a Steady-State Economy* maintained that economic growth for growth's sake is destructive. Investigative reporter Jack Shepherd penned *The Forest Killers,* an emotional documentary accusing the United States Forest Service of despoiling national timber and leaving an "empty legacy of a squandered resource."

The exhortations against overconsumption have had little impact on politically constrained policy makers who must balance jobs for voters as well as preserving the environment. In this country, the GNP (Gross National Product) is the barometer of a successful economy. The GNP is expected to grow every year; GNP growths considered small often trigger remedial federal government programs. To prevent inflation, the Federal Reserve Board manipulates interest rates to stimulate or restrain growth. However, politically, United States policies have not reconciled preserving the environment with providing jobs. The jobs versus the preservation of the environment argument persists. The industrial sector and many communities maintain that halting the resource extraction or utilization on which they depend for jobs to preserve the environment deprives them of their livelihood. The environmental community maintains that, on the contrary, good environmental performance helps the economy grow. Both views have demonstrated merit, and the controversy continues.

Sustained economic growth has always been a welcome signal for the forest industry; it means more homes using lumber and wood furnishings, increased industrial activity, and more disposable income for crafts and gifts. The trade journals of each segment of the industry carry growth statistics. Think-tank policy meetings in Washington, D.C., urge curtailing growth while a forest industry trade association meets in Atlanta to plan how to grow a larger market for its products. Thus, the industry and government policymakers are at odds.

--- · -- · -- · -- · -- · -- · -- · -- · -- · -- · -- · -- · -- · -- · -- · -- · -- · --

Earth Day: Past, Present and Future

The Past as Prologue: The original Earth Day in 1970 was a huge success, quickly raising public visibility for environmental concerns. In *public opinion,* among *environmental organizations,* and in *legislation* to protect the environment, the first Earth Day had notable influence, some of which is illustrated below.
   – Bill Lunch, OSU Political Science

Public Opinion: Earth Day focused attention on environmental issues and the public responded in 1970, but concern about the environment, particularly pollution, had been growing for some time. For example, consider the increases in concern about air and water pollution shown in the following public opinion findings:

• • • • •

   Respondents viewing pollution as "somewhat" or "very" serious (in %):

	1965	1966	1968	1970
Air Pollution	28%	48%	55%	69%
Water Pollution	35%	49%	58%	74%

   – from reports by the National Opinion Research Center (NORC)

• • • • •

Environmental Organizations: The increased awareness of environmental problems stimulated by the first Earth Day produced two related developments: first, established environmental organizations enjoyed a surge in membership; and second, a number of new environmental organizations were created. Here are some examples:

Number of Members (in Thousands) in:					
Organization	Founded	1960	1969	1972	1990
Sierra Club	1892	15	83	136	560
Audubon Society	1905	32	120	231	600
Wilderness Society	1935	10	44	51	370
Friends of the Earth	1969	–	–	8	30
Natural Resources Defense Council	1970	–	–	6	168

• • • • •

Legislation: The congressional response to the first Earth Day was quite strong. A number of landmark environmental protection laws were passed within a relatively short time. Some of those are shown below:

Year	Law
1969	National Environmental Policy Act (NEPA)
1970	Clean Air Act
1972	Water Pollution Control Act
1972	Marine Mammals Protection Act
1972	Federal Insecticide, Fungicide, and Rodenticide Act (FIFRA)
1973	Endangered Species Act (ESA)
1974	Safe Drinking Water Act

ENVIRONMENTAL STRATEGIES WORK

The strategies used by environmental pioneers — those who recognized and embraced environmental ideas early — were time tested. First, they raised the consciousness of the American people on the value of the environment. Their movement astutely focused on obtaining a wide base of support: schoolchildren, idealists, the disgruntled and cynical, the growing number of educated Americans with leisure to worry about the planet, the tidal wave of baby boomers entering their adventurous twenties and thirties, a media hungry for news, and persons genuinely worried by the prophesies of doom.

As in most movements, action was the key to visibility. Few industrial giants have ever organized a performance as stunning as the 1970 Earth Day. All the pieces were orchestrated: coordination; forming alliances; collaboration; visible actions; critical masses; a noble cause; and organizing dynamic, enthusiastic young collegiate men and women. Another important factor was that, in an era when women were still struggling to enter the professional workforce and break through the glass ceiling, women found themselves welcomed as part of the environmental crusade.

A crucial key to the movement's success was that the media was willingly co-opted to report and support the uplifting message of saving the world. Traditionally, stories that attract readers rely on conflict between good and evil. The media quickly identified the villain as industry and the heroes as passionate environmentalists.

Environmental leaders early recognized the difficulty of maintaining the crusading zeal that energized the movement in its early days. However, the federal government, in its subsequent efforts to modify the impact of environmental regulations, has repeatedly provided the spurs to sustaining the unbridled enthusiasm. President Ronald Reagan's appointment in late 1980 of James Watt as Secretary of the Department of the Interior proved one of these spurs. Before his appointment to his position, Watt's Mountain States Legal Foundation had taken on the Environmental Protection Agency, the Sierra Club, the

Environmental Defense Fund, and the Department of the Interior, and won often enough to be tagged as an enemy of the environment. James Watt's programs infused new energy into the environmental movement.

In the 1990s, the young Republican newcomers to Congress provided another spur with the passage of the "salvage rider" to the timber salvage bill. The salvage rider injected new vigor into the environmental troops. "Congress energizes environmentalists" headlined the Knight-Ridder Tribune Service in a March 1996 article. Niki Stevens, executive director of a newly formed environmental group, Restore America's Estuaries, was quoted as saying that Congress is doing them a big favor. "They're mobilizing people who care about the environment." At the same time, the giant environmental groups like the Sierra Club and the Wilderness Society, enjoyed a growth, or at least a stabilization, of their membership.

> It was supposed to be a good thing, and everybody was supposed to be happy. But in the nine months since it became law, the timber-salvage rider, which was part of the budget President Clinton approved last July, has created intense acrimony on all sides. And the bill has only been partially carried out.
>
> The salvage rider was supposed to free 1.5 billion board feet of dead or diseased trees for cutting. It also released old sales that have been held up by environmental reviews, and suspended environmental reviews, and suspended environmental laws, including the Clean Air Act and Endangered Species Act, with regard to the salvage cutting.
>
> But timber groups now complain that the bill isn't being implemented. In Congress, there have been multiple attempts to modify the bill. Environmental groups have joined with Native American tribes and fishing organiza-

tions in condemning the harvest of old-growth green trees in national forests — particularly in the Pacific Northwest — which were released under the rider.

So far, the primary effect of the salvage rider has been to generate a fever-pitch of negative public opinion....

The rider has become a particular embarrassment for President Clinton, who has tried to build a reputation as environmentalist....

Environmentalists' opinions of the salvage law have been blunt. "In one word, it's a disaster," Jim Jontz, head of the Western Ancient Forest Campaign, told the *Los Angeles Times*. "Everything we've learned about how not to log in the West has been thrown out the window. This is the old-style, ugly clear-cuts the Forest Service said they'd never do again."

While environmentalists decry the effects of the salvage rider, timber groups like the Independent Forest Products Association (IFPA) in Portland, Oregon, are arguing for the salvage law's full implementation....
(*National Home Center News*, April 15, 1996)

Again, as in the early days of the movement, the environmental advocates accomplished their mission. Poll after poll shows strong support for environmental protection. Some examples:

Luntz Research, March 1995, reported that 62 percent of Americans want Congress to make environmental protection a priority over cutting regulations.

A June 1995 Harris poll showed that, two to one, Americans do not favor spending reductions on the environment.

A Gallup *CNN/USA Today* poll in April 1995 showed that twice as many Americans believe that protection of the environment should be given priority over economic development.

THE RELIGION OF THE ENVIRONMENT

In the 1990s, man's dominion over the earth received a final blow from Biblical scholars. Revisiting the passage in Genesis, they determined that God's command really meant nurturing the earth rather than dominating it. In place of conquering nature, the environmental view called on man to live in harmony with the earth.

Since the 1970's, grumblings that "environmentalism is like a religion," which was often intended as a disparaging comparison, have clouded the discussion. Recognizing that the Endangered Species Legislation was under reconsideration, a conference of scientists and religious leaders maintained that, "Helping people understand that wild salmon — or any other creature — have spiritual value is essential to saving endangered species."

At first glance, the Northwest's disappearing salmon runs seem to have little to do with religious faith of any kind.

Look again, say a group of Oregon State University scientists and campus religious leaders.

Helping people understand that wild salmon — or any other creature — have spiritual value is essential to saving endangered species, said Donna King, campus minister at Luther House.

"Our intent is to help people find a religious basis to be involved with the earth," King said. "There's a lot of environmental problems out there, and we hope to get church people involved."

Today's conference, "Swim with the Salmon: Ecology of Faith, Creation of Hope," may be a first step in that direction for a lot of local people, King said. (*Gazette Times*, May 20, 1995)

_._

In the fall of 1995, the Ecumenical Patriarch Bartholomew I, head of the Greek Orthodox Church; and His Royal Highness The Duke of Edinburgh, called a conference attended by scientists and representatives of numerous religions. The group reached consensus on a new definition of the Apocalypse. They decided that the Apocalypse should be viewed not as a vision of inevitable doom, but as a prophesy of events which might be avoided if humans take their responsibility for stewardship of the earth more seriously. The Greek Orthodox Church issued the revolutionary pronouncement that it will recognize an expanded definition of sin that includes "sins against nature" that degrade the environment. The meeting concluded with a commitment to a religious responsibility for safeguarding the environment.

A NEW CULTURE EMERGES
IN INDUSTRIALIZED NATIONS

The environmental movement prodded the American conscience, reflecting the views of of philosopher Henry David Thoreau: "If a man walks in the woods for love of them ... for half his days," he wrote, "he is esteemed a loafer; but if he spends his whole day as a speculator, shearing off those woods, he is esteemed industrious and enterprising — making Earth bald before its time."

Writing in the 1800s of his beloved refuge, he blasted progress and civilization. Thoreau was one of the first to express the concepts of the environmental movement: "Nowadays, " he wrote, "almost all of man's improvements, so called, as the building of houses, and the cutting down of the forest and of all large trees, simply deform the landscape...."

Who could quarrel with cherishing the denizens of the forest? The "Bambi syndrome," so called because of the popular Disney movie, *Bambi,* was engraved in the hearts of Americans. Few would have disputed these aims if the environmental movement had simultaneously recognized and provided for the requirements of civilized people to survive by earning a living rather than spending "half his days loving the woods."

The environmental movement proved a catalyst for profound and fundamental changes that has experts still struggling for definitions. Human resource expert Arnold Deutsch termed it "The Human Resources Revolution." In 1976, Ian Wilson of General Electric Corporation observed: "We in the United States are in the throes of a major transition comparable, say, to the transition that took place when man the hunter and nomad became man the settled farmer; ... when civilization — in its literal sense — became the norm.... Now our society is on the brink of becoming something different, a new form of society.... Whatever future generations call it, it will be radically different from our present world ... in its attitudes, its values, and its institutions."

Wilson's prophecy proved eerily correct. Caring for the planet expanded into the demand for new social and economic rules of behavior. By the early 1990s, the passage of the environmental acts enacted almost thirty years earlier had changed industries, society, and values. A survey for *Psychology Today,* taken in 1964, showed that Americans were mainly concerned about a better or decent standard of living; the survey didn't mention the environment. Since then, two generations of schoolchildren have learned that saving the planet is more important than saving the economy.

_·_

During this time recycling was transformed from a Boy Scout and Girl Scout money-raising project to a booming business. Recycling newspapers and bottles and cans made proverbial silk purses out of sows' ears. Thrift became the vogue. In the 1980s, educated, higher-income families fashionably adopted recycling to save the planet. By the mid-1990s, recycling has achieved the status of an environmental product attribute. Stationery and packaging sport mandatory "made of recycled material" slogans. Although the United States has not adopted policies as strict as Germany in mandatory recycling, it has become the norm.

Along with the recycling trend, "natural" materials became more fashionable than man-made goods. Natural saved the earth, while man-made used its resources. It might reasonably be expected that natural would heighten the appreciation of products from the woods. But, by the mid 1990s, Americans were told to help the planet by not using wood and thereby sparing trees. Recycling bins appeared in offices to reuse papers "to save trees." Packaging adopted the "less is more" slogan to encourage minimizing packaging to save trees. Blowdryers installed in public rest rooms were preferred to paper towels "to save trees."

"Reused" wood also gained popularity in the wood fashion world. Urban wood, wood salvaged from demolition projects, or any wood that had an earlier life was reincarnated as a viable means of saving the forests. Labels boasted that products were manufactured from reused wood.

The word *green* soon was equated with good for the environment. *The Green Reporter* reported that hotels practicing environmental savings such as not washing linen and towels every day for guests remaining in the same room qualified for "green suites." "Green reports" in consumer magazines, "green marketing" and other "greens" won consumer acceptance for saving the environment.

THE FOREST INDUSTRY REACTS — OR MOSTLY DOESN'T

The forest industry, struggling to survive under the old rules and old ideas of success, barely noticed that a new idea was taking root in the society it served. And when the baby boomers embraced these radical new environmental ideas, industry airily dismissed them as youthful fervor and lack of understanding, a passing fad. "These kids will grow up, cut their hair, and shave their beards. We'll have a little recession, and everyone will forget about this nonsense," was the misguided conclusion of one industry leader.

The industry never imagined that their public image could become so tarnished. After all, they thought, wood is a renewable resource and essential to life. Do we have any choice but to grow and harvest timber? Don't worry, trees grow!

So the scholars worrying about the future of the earth heard relatively little from the forest industry while they were introducing the new ideas and legislation that would change the forest industry.

Environmental requirements barely rippled the forest industry — until the late 1970s. And then it was often under the banner of responsiveness — a continuing awareness of public affairs — consideration of ethical behavior in each situation. Companies assessed their responsiveness by how much and what kind of response they made. The outside community, however, assessed responsiveness according to company actions and the consequences of these actions.

IMPACT OF THE ENVIRONMENTAL MOVEMENT

If they thought about it at all, the men who sweated or froze in the woods or fretted in their offices about the price of logs, the green chain pullers, the hewers and haulers, had other concerns besides counting the number of people on the fragile planet. When they finally noticed the changes that had been wrought by environmental activism, the rank and file of the industry were bewildered when suddenly cast in the role of environmental villains.

At first, the forest industry indulged in denial, then they adapted. The larger companies and trade associations attempted to protect their flanks by creating a new profession — government affairs specialists — to explain the forest industry's position to state legislators and Congress.

However, problems arose when social demands clashed with the traditional and necessary objectives of the companies — producing products to meet people's needs, assuring sufficient profits to secure financing, fostering growth to remain competitive, and obtaining a favorable image to maintain credibility.

The impact of the environmental movement is largely a generational phenomenon. Men and women in the United States under fifty were raised to believe that caring for the environment is the norm, a "given." To the senior and upper-middle-aged executives in the forest and other industries, environmental regulations and the demands of the current environmental culture have forced a sea change in the way they must conduct their business. Their foot-dragging response to these changes reflects the attitude of the era in which they were raised. As the younger generation, schooled in caring for the environment, assume management authority, the responses of their industries may be expected to be more attuned to the environmental culture.

Chapter 3

Conflicts Take Center Stage

"...there is no inherent conflict between achievement of reasonable economic goals and reasonable environmental goals," declared John B. Crowell Jr. in a speech in 1983. Crowell, then Assistant Secretary of Agriculture for Natural Resources and Environment, added, "We are now certain national forest plans can more adequately explore the untapped management opportunities that exist for increasing benefits to the public both for commodities and amenities."

Was anybody listening to these optimistic statements? In 1996, members of the Sierra Club, for the first time in its 100-year history, voted by a two-to-one margin to ban commercial logging in national forests. The vote, according to a carpenter's union spokesman, cost "any credibility [the Sierra Club] once had." Commenting on the action, an American Forest and Paper Association (AF&PA) spokesman declared: "The club has taken itself out of the debate on how to manage our forests.... Since its position is unalterably opposed to timber harvesting, it no longer can be considered to be interested in an equitable solution to forest problems."

Earlier that year, demonstrators protesting a forest industry meeting in the State of Washington waved signs bearing the slogan, "Zero cut on public lands." Washington Senator Slade Gorton, considered the "enemy" by mainline environmental groups, commented on the demonstrators' signs: "When you see the real message of the other side is zero cut on public lands, that is by definition an extremist position, as extreme as you can get, with no consideration of people or the communities they live in."

These comments illustrate the conflicts in which the forest industry has been engaged for the past thirty years. The conflicts are inherent over the use of what may be called the most diversely valuable land in the United States. The conflicts persist, and few measures have been devised to prevent, avoid, or minimize continuing conflicts.

Many of the conflicts originated over the management of the publicly owned National Forest System. The public owns 17 percent of the total productive forest land in the country, but 65 percent of the softwood forests are in the Western states, more than any other section. National forests capture the attention of Westerners as the mainstays of rural economies where logging, milling, and processing timber are the major source of employment and community income.

The 155 national forests and nineteen national grass-lands, comprising slightly over 8 percent of the area of the United States, loom large on the national scene: They provide forage for three-and-a-half million cattle, homes for half the country's big-game animals, and contain the headwaters of a large part of our drinking and irrigation water. More than 40 percent of outdoor recreation on federal land and almost a third of the National Wilderness Preservation system occurs on this 8 percent. More than half the softwood timber in the country is also located in national forests.

However, the rest of the country cannot casually dismiss anxieties about managing the national forests as only of regional interest. The national forest makes an easy target for environmental actions, but federal policies on managing these forests filter down into the state and private forest policies. Insects such as the southern pine beetle can devastate entire forests but can't read boundary maps, and fly freely from federal lands to private lands. In only one year, 1994 to 1995, the southern pine beetle increased fivefold, invading four times the number of acres in Arkansas, Louisiana, Mississippi, Alabama, Florida, Georgia, and North and South Carolina, costing more than $300 million in damages. The original National Forest Organic Act of 1897 that forged the principle of managing the "public timber reservations" that Congress had created six years earlier provided for sustained yield of timber mixed with multiple use of the forest. The Organic Act directed that the reserves be managed to improve the forest reservations, aid in water flow, and assist in producing timber. The timber from the reserves — which later became the national forests — was sold by the General Land Office to lumber companies and the railroads.

The multiple-use provision of the Organic Act proved to be the Pandora's Box in managing the national forests.

First out of the box was the request by a trio of hunters in 1972 to the Izaak Walton League, an advocate of hunting, to sue the Forest Service for permitting clearcutting of timber on West Virginia's Monongahela National Forest, claiming that clearcutting was driving game animals out of the forest. The courts ruled that clearcutting violated the multiple-use provisions of the Act and declared that the only future sales on the Monongahela permitted were for the harvest of "dead, matured, or large-growth trees." The effect of the ruling was to prohibit clearcutting in national forests, sending the U.S. Forest Service and the forest industry into a tizzy of uncertainty.

Timber sales were slashed across the national forests until 1974, when Congress passed the Resources Planning Act (RPA), which required the development of fifty-year plans for the national forests, with updates every five years. Two years later, Congress contributed the National Forest Management Act (NFMA), which specifically allowed clearcutting in appropriate places in the national forests and required every national forest to appoint an interdisciplinary team (IDT) to develop a management plan giving equal weight to all forest resources.

The NFMA planning was greeted as a great idea by both the forest industry and environmental advocates. The industry thought it would settle the clearcutting issue and bring stability to the forest. The environmental advocates were pleased that other resources would have official standing in forest management. For the U.S. Forest Service the IDT proved another unsettling development. In 1958, 90 percent of the Service workforce consisted of trained foresters. Passage of the RPA and NFMA required biologists of all kinds, social scientists, soils and water specialists, transportation specialists, and an array of other specialists. The percentage of trained foresters dropped to 50 percent in 1973 and continued to drop. Forest management was further complicated by passage of the Endangered Species Act in 1973. The net result of all the planning skyrocketed the cost of timber sales.

THE FOREST INDUSTRY'S VULNERABILITY TO CONFLICT

Forests are inherently vulnerable to conflict. The varied stakeholders with their many interests guarantee a conflict about any proposed entry into a forest, whether by chainsaws or skis or homes. Forest-related conflicts fall into five groups, keeping the forest industry hopping from one issue to another. Some conflicts encompass all five types like a bear hugging a victim:

- Conflicts about using the space occupied by forests — housing; resort and recreation developers; road building; dam construction.
- Conflicts between competing users of the forests, in which the forest industry is often caught in the crossfire — hunters versus campers, ornithologists versus foresters, fishermen versus loggers, snow-mobilers versus skiers; homeowners' aesthetic requirements versus private property rights of forest landowners; and Native American cultural values versus logging.
- Conflicts between environmental values and forest industry practices, such as clearcutting, use of herbi-cides, spiritual aspects of old-growth forests; role of endangered species; timber harvesting as anti-nature; and exploitation of rainforests.
- Conflicts between social and economic equities and entitlements and the forest industry — the public's

assumed entitlements to clean air and water; a livable climate (global warming), and freedom from the effects of toxic pollutants from processing wood for paper or products; conflicts between the economic contribution of the forest industry to communities and environmental values; protecting the forest ecosystem or endangered species versus jobs; forest fire management versus the economic benefits of tourism; log exports versus community jobs; social/political conflicts — control of public and private forests; large versus small forest industries; and responsibility of forest industry to society.

• Conflicts between the human-centered (anthropocentric) view that natural resources are put on earth for the benefit of man, and the ecocentric view that all parts of nature are equal. The ecocentric view brings trees into the legal system as holders of legal rights. Harvesting trees can be considered as depriving trees of their legal rights. (That trees should have standing in the law was endorsed by Supreme Court Justice William O. Douglas and supported by two other justices.)

CONFLICTS ABOUT THE LAND OCCUPIED BY FORESTS

Forests occupy 700 million acres, about a third of the United States' land area. Cropland, much of it carved out of forests, occupies another 400 million acres. These 1,100 million acres, about half of the United States, are valuable acres with multiple uses.

Space was not an issue in 1790, when there were fewer than five people per square mile. With the 1803

Louisiana Purchase, the United States had more space in the public domain than it could manage. As the population increased sevenfold by 1850, the national interest called for transferring land out of the public domain to private ownership. The resulting privatization of western lands was the largest ever undertaken in the history of the world.

The importance of land grew with the population. By 1970, the pivotal year of the environmental movement shift, the population had expanded to almost sixty people per square mile. Twenty years later, it had expanded again to seventy people per square mile.

As the population continued to grow, the space occupied by forests increased in value. People with diverse interests began to quarrel over forest space, often engaging attorneys to represent their positions. Today, almost any taking of forest land guarantees a fight, especially on publicly owned lands.

The plan to construct a multi-million-dollar ski resort in Pelican Butte, located in the Winema National Forest northwest of the town of Klamath Falls in Southern Oregon, is beset by all five types of conflicts. The lust for space pitches competing interests, environmental values, and social and economic equities against each other. The competing interests were required to spend millions of dollars for environmental impact statements, permits, experts, litigation, hearings, and other hurdles to have their requests considered.

> A Southern Oregon company with a long record of commercial success is drawing up plans for a multi-million-dollar destination ski resort. The Pelican Butte project would be in the Winema National Forest northwest of Klamath Falls.

The $30.7 million plan must clear environmental hurdles, but plans call for 39 runs on 612 skiable acres with a gondola lift, four chairlifts, a T-bar at the summit, Nordic ski trails, snowmaking machines, an 18,000-square-foot base lodge, and 30,000-square-foot midmountain lodge.

The ski area is intended to attract the high end of the market, drawing two-thirds of skiers from outside Southern Oregon. It's designed to handle 5,560 skiers a day, with one high-speed detachable quad, two fixed quads, and one double chair....

"It's a good concept for a high-quality, environmentally sensitive regional ski area," said Sheridan Atkinson, the project manager. "There's an alignment: a quality developer, a region that could use the economic impact, and a Forest Service that is at least reasonably receptive at this point."

Pelican Butte would be developed as what promises to be "the most environmentally driven ski area to date," said Harold Derrah....

With most of Oregon and Washington's undeveloped ski potential resting on national forest land, the red tape is enough to scare off some developers.

The required environmental assessments are time consuming and expensive; and once complete, they may prove the project infeasible.

A key to gaining federal approval lies in illustrating need. The backers of Pelican Butte see plenty of potential in nearby residents, Californians, who now trek to Mount Bachelor, and guests at the sparkling new destination resort. (*The Oregonian,* April 5, 1996)

Pelican Butte is planned as "the most environmentally driven ski area to date," according to the developers. The ski area, with thirty-nine runs on 612 skiable acres, and two mountain lodges, has the economic interests in the region salivating with the prospect of attracting affluent tourists and providing jobs. But economic feasibility takes second place to the other requirements when a forest is invaded.

The United States Forest Service's obligation is to "balance the competing interests." The cast of characters in this cliffhanger includes, in addition to the developer and the Forest Service:

A bald eagle researcher
The city of Klamath Falls
Klamath Falls business leaders
Former timber workers in the now-closed sawmills
The Natural Resources Council
The endangered spotted owls
An environmental consulting firm
A wetlands investigation
A ski resort and design firm
An attorney specializing in environment
Attorneys not specializing in the environment
The Klamath Indian Tribe
Developers with an interest in another, nearby, ski resort
The courts
Skiers

Almost $3 million was spent in environmental investigations alone in the plan to attract almost 6,000 skiers a day. Controversy even existed over the data on the dependability of the snow cover. The experts could not agree on the how much snow skiers could anticipate:

The Forest Service found a lack of dependable snow under 6,000 feet.

University experts found an adequate eighteen inches of snow at 5,600 feet.

The developers maintained that there are thirty-six to forty-eight inches of snow at 5,200 feet.

The environmental groups and attorneys say, "There's no way around the impacts of a ski area." A field representative of the Natural Resources Council warned: "If the developers move forward, they should expect a long and expensive battle with the environmental community."

The developers maintain that the area is perfect for skiers and will be developed with environmental sensitivity. Skiers will ultimately answer the question, "If you build it, will they come?"

Klamath Falls officials, businesses, and unemployed residents say, "We need the jobs."

The forest biologists say, "The project would destroy thirty-two acres of old growth, and affect sixty-nine acres of suitable owl habitat, and sixty-seven acres of eagle habitat."

The Klamath Indian Tribes believe that the entire Winema Mountain is a religious site, with archaeological and spiritual areas and vision-quest sites that should be preserved.

The outcome is that habitat, religion, jobs, and recreation all compete for the same forest. The right to log timber in that forest had previously been stripped from the forest industry by legislation. It is no longer a significant player in this scenario.

As the 1990s began, the industry was numb with the stress of conflicts. A manager of one of the country's large papermills concluded, "There's no way we can please anyone. So many people think their use of the forest is the most important, we're always caught in an argument with someone."

Conflict is inevitable for an industry that plays so many roles and serves so many stakeholders with diverse interests. But today there is more fuel for controversy than ever before.

News headlines tell the story of the quest for space for the increasing population, which is the origin of many conflicts. About a third of America's land area is covered by forests today, in contrast to about 46 percent in 1600. Some of the original forest was cleared for agriculture; consequently, the 54 percent of non-forested land in 1600 has increased to about 68 percent today. About 15 percent of the original forested acres have been cleared to grow food, and to build dams, roads, and developments for the growing population. The doubling of the United States' population in the past fifty years has exerted unprecedented pressure for the forest land.

A rapid growth in lumber production in the 1800s raised fears that today might sound like deja-vu. Douglas MacCleery of the U.S. Forest Service, in his 1992 *American Forests,* published by the Forest History Society, describes the concerns of that period: "… about future timber supplies … implications for increased flooding and watershed damage, declining wildlife populations, harm to the beauty of the American landscape, and even concerns about how forest clearing was affecting the climate itself." The fears in the 1800s included logging's adverse effects on watersheds and other environmental values. Predictions of a coming timber famine led to calls for action to preserve forests. Sounds familiar!

The United States' population has more than tripled since 1900, and our standard of living has increased beyond our grandparents' dreams. In addition to population pressures, improved transportation, shortened working days and increased leisure, bigger incomes, urbanization, and a longing for communion with nature have combined to heighten an awareness of forest activities.

As the forest industry adopted new forest management methods — such as clearcutting, pesticides, fertilization, building logging roads, and burning — in response to the post-World War II demand for timber, for the first time the public witnessed the visible changes to the forest. And they didn't like what they saw happening.

Added to this awareness, in the 1950s, affluence and mobility combined to awaken America's interest in outdoor recreation opportunities in its national forests. Millions of families visited the forests for recreation — ninety-three million by 1960, a vast increase from the sixteen million immediately after the war.

The Multiple-Use Sustained Yield Act in 1960 marked the transition from the focus on timber production to multiple-use forestry in publicly owned forests. Outdoor recreation, protection of watersheds, and wildlife, gained equal status with timber and grazing. Four years later, the passage of the Wilderness Act set aside wilderness areas, protected from timber harvesting. Despite contentious debate, by 1990, over thirty-three million acres in national forests had been preserved from timber cutting.

In 1976, The National Forest Management Act (NFMA) set detailed guidelines for managing national forest land, requiring citizen participation in decision-making. Congress released broad mandates and detailed guidelines for Forest Service decisions in local situations.

These acts inserted public opinion (and therefore public perception) into national forest planning. The public voice heard in national forest plans proved sufficiently loud to carry over into state forest management, and into most of the remainder of the forest land in the United States — the non-industrial private forests (NIPF), owned predominantly by small companies and ten million individual landowners.

As evidence of the impact of this legislation, today we talk about "zero cut on public lands."

Actions on public lands have been the bellwethers for the future of private forests. At first, the extremists' skill in using tactics that captured media attention lulled the forest industry into complacency about objections to its practices. "It's only the extremists talking," they declared, "not the general public." However, the many surveys conducted on public perception fail to support the belief that "it is only the extremists talking." Recognizing the influence of "those extremists" will go a long way in helping the forest industry regain its status as a respected environmentally acceptable industry.

The continuing controversies pit segments of the forest industry against each other and tear at the fabric of society. Exporting logs frays relationships between the timber industry, labor unions, sawmill workers and mill owners. Forest-based rural communities, faced with unemployment when timber has been withdrawn from harvest, battle with urban sophisticates, who consider forests their recreational playgrounds.

Certification of sustainable forest management is involved in the perennial fight for market share between large and small companies, and the "haves" and "have-nots" of the industry.

Hunters collide with industry and campers over their place in the forest. Contention rages between the timber industry and farmers/ranchers; between the timber industry and recreation interests, between the timber industry and

environmental organizations, between the timber industry
and tourism, between the timber industry and population
pressures encroaching on its boundaries. Fish, wildlife,
water resources, air and other natural resources compete
with timber industry practices. Two-and-a-half decades of
environmental education have put boxing gloves on
children willing to fight objectionable forest management
practices.

Mountainous Dark Divide Will be Bridged in Court

TACOMA — The Dark Divide is anything but dark,
except for the harsh words between bikers and hikers.

For twelve miles the mile-high ridge extends through
the Gifford Pinchot National Forest, rich with blue lupine
and white lilies, offering views of mounts Rainier, St.
Helens, Adams and Hood. Features such as 5,238-foot
Dark Mountain and Dark Meadows are named for John
Dark, a nineteenth-century gold prospector and speculator.

Trails in the 55,000-acre roadless area west of Mount
Adams in southwest Washington are open to motorcycles
under the forest's management plan but are too primitive
for all but the most accomplished riders.

In 1994, at the urging of backwoods enthusiasts in the
Northwest Motorcycle Association, U.S. Forest Service
officials approved $150,000 to improve some of the trails.

Two months ago, acting on a lawsuit filed by the
Washington Trails Association and eleven environmental
groups, U.S. District Judge Barbara Rothstein rejected the
plan. She ruled it was adopted "arbitrarily and capri-
ciously" and that hikers' concerns were "either ignored
totally or brushed aside summarily."

...Ira Spring of Edmonds, author of more than three
dozen regional hiking guides and co-founder of the trails
association, said dirt bikes have left steep ruts and new
trails as riders departed from established paths to avoid
snow or old grooves.

------- -- -- -- -- -- -- -- -- -- -- -- -- -- -- -- -- -- -- --

"They're noisy, they're fast and they smell horrible when you're huffing and puffing behind one," Spring added. "There's no such thing as multiple use."

The hikers are "absolute zealots," like "little kids who have learned they can get their way if they scream loud enough," said Dave Hiatt, 48, a motorcycle association activist and vice-president of Hiatt Pontiac in Tacoma. (Associated Press, September 12, 1996)

THE GLOBAL ENVIRONMENTAL THRUST

Concern about tropical rainforests gave birth to the Woodworkers Alliance for Rainforest Protection (WARP) in 1989, changed to the Good Wood Alliance in 1995. Scott Landis, editor of their journal, *Understory,* relates in his Winter 1995 editorial how WARP emerged from his visit to the Peruvian Cooperative for tropical forest management. WARP began life as "a diverse group ... [that] wanted to understand the real threats to forest management and the role that wood users might play in addressing them."

As other organizations became involved in carrying the torch, Landis says, "We ... focused our gaze on temperate and boreal, as well as tropical forests, and we have cast our net more broadly to attract the architects and designers, manufacturers and retail businesses...."

Conflict between the forest industry and environmental advocates is not peculiarly American. Conflicts have escalated in the past forty to fifty years in Sweden, Germany, France, Norway, and Great Britain. Although the United States' standing volume of trees is seven times that of Sweden, and close to twenty times the standing timber compared to Germany, France, Finland, and Norway, the struggles are remarkably similar: intensification of forestry, increased recreational use of the forests, growth of the environmental movement, and the stronger role of the media. The public was simultaneously receiving the same message in Swedish, or German, or Finnish, or English.

In all these countries forests became the focus of struggles between the many interest groups. A European Forest Institute Report, *Forestry Conflicts from the 1950s to 1983,* printed in Finland, concludes:

"The factors that led to forestry conflicts were basically not caused by mistakes made in forestry practices, by the lack of information, by the bad will of the critics, or by exaggeration by the media, but were integral parts of the social developments of industrialized countries."

The report further confirms the conclusions of many Americans:

"Forestry professionals in all the countries studied made an incorrect judgment that with efficient, correct information and propaganda, the criticism would decrease — a judgment still being made today. It seemed that the forestry profession was so tied to its tradition ... that it could not see the social development in which forests received wholly new values and uses. Instead, these changes were [experienced] as unnecessary and as a threat to the traditional status."

MORE FOOD FOR ARGUMENT
THAN FOR THOUGHT

There is no shortage of conflicts to sour the relations between the stakeholder public and the forest industry. Public participation, mediation, political manipulation, conferences, "summits," public hearings, public relations, and education, have had questionable results in the struggle to find that elusive "balance" between conflicting ideas. Yet, after years of seeking answers, the basic controversies remain largely unresolved. The word "risk" is batted back and forth in conflicts, as if risk was quantifiable and evoked the same emotions in all people. The absence of single definitions for risk introduces a considerable risk in arriving at conclusions. So risky is discussing risk that a publication called *The Risk Advisory* is available to diminish the risk of a mistake.

A catalog of the controversies gleaned from media headlines reflects specific problems that surfaced in the last twenty to thirty years. Some of these are listed, without prioritizing or categorizing:

Controlling pollution

Protecting endangered ecosystems

Protecting wilderness and roadless areas

Protecting private property rights ("wise use")

Protecting public rights

Protecting endangered species

Protecting water resources

Protecting the right to litigate

"Rights" to natural resources

Saving diminishing natural resources

Saving rainforests — tropical and temperate

Saving old growth

Saving endangered fish

Saving the ozone layer

Saving jobs

Slowing global warming

Eliminating pesticides and herbicides

Protecting against sick houses

Harvesting timber (logging)

Salvage logging

Clearcutting
Preserving biodiversity
Preserving recreation opportunities
Maintaining forest health
Maintaining community employment
Maintaining clean air and clean water
Managing forests for sustainability
Facilitating sustainable development
Public oversight of forest management
Public participation in forestry decisions
Sparing the deer and other animals — "Bambi Syndrome"
Avoiding depletion of forests
Avoiding plantations
Responding to population pressures
Exporting logs
Protecting the environment by recycling
Managing national forests
Reducing forest fires
Federal versus state versus private control of forests
Large versus small forests
The role of the United States Forest Service
Using trees to make paper
Exploiting the forest
Native American-white conflicts over fishing resources

Pariahs in forest uses:
Locally Unwanted Land Uses (LULUs)
Temporarily Obsolete Abandoned Derelict Sites (TOADS)

None of these controversies has yielded to definitive solutions. Individual judgment, personal values and objectives determine the "right" answer. The answer that prevails is decided more by the strength and skill of the constituencies developed to promote each side's idea of right than scientific findings.

Vice President Albert Gore, addressing the 1996 annual meeting of the American Association of the Advancement of Science, summed up the dilemma of determining the right answer:

"We continue to rely on a know-nothing society ... a society in which the spigots of discovery are twisted and turned off.... This know-nothing society bases regulations on suspicion instead of science ... and claims that global warming is the empirical equivalent of the Easter Bunny."

Arguments over harvesting timber range around curtailing logging to save old growth, rainforests, endangered species, and forests that furnish pulpwood. Ecological values spawn debates about ecosystems, biodiversity, roadless areas, and endangered species. Social priorities create heated arguments over protecting water resources, the ozone layer, and recreation opportunities. Public policies shape their own debates about property rights, global warming, jobs, equities, and the role of the United States Forest Service.

The public influences forest management, usually based on the media reports. Newspaper headlines, magazine articles, television programs, litigation, and public hearings infrequently complain about wood products themselves, unless a failure — such as experienced with a defective batch of Oriented Strand Board (OSB) — piques the ire of the public. The important role of forests for shelter doesn't rate headlines except when the price of lumber increases.

Conflicts may be regionally limited. In the Pacific and

Mountain states, where more than half the forest land is in the national domain, management of national forests is a persistent, major concern.

Some issues worry scientists more than they do the public: How many Americans understand biodiversity, forest health, ecosystem management, and "old growth"? Extensive media coverage has pushed these scientists' concerns into public awareness. Private property rights loom as a major issue as non-industrial private forests (NIPF) come under more stringent regulations.

Does the ongoing debate affect people's opinion of the forest industry? "Yes," answer many CEOs. The AF&PA Sustainable Forestry Initiative, which is a move to demonstration of sustainable forest management, and the current changes in forestry practice, are attempts to diffuse the controversies and instead center attention on the positive impact of the forest industry.

There are still other industry executives who shrug off the debate as irrelevant and unrelated to how they run their businesses. "We believe in the free enterprise system, and we believe we have a right to manage our business the way we consider best," was the response of a top executive of a major timber company when questioned about exporting logs.

"We can't react to every [crackpot] idea that comes along," explains still another executive. It is apparent that opinions on many of these difficult issues reflect generational differences with those under forty or fifty years old more tuned to the environmental culture.

Of the almost fifty controversies surrounding the forestry industry, intense current media attention centers on seven: clearcutting, old growth, herbicides and pesticides, endangered species, rainforests, jobs, and litigation. Of course, these seven issues involve other controversies and concerns. The bottom line is that almost all the controversies except for two or three center on forest industry practices, with the forest industry as the defendant.

THE CLEARCUTTING CONTROVERSY

In the clearcutting controversy environmental and aesthetic values directly oppose a forest industry practice, a practice that riles the public probably more than any other industry action. Even Horace Greeley, former chief of the U.S. Forest Service, accused by the environmental community of being married to the timber industry, described a clearcut as "a fitting approach to Dante's Inferno."

Companies Defend Clearcutting

...Plum Creek, once referred to as the "Darth Vader" of the industry, is among the nation's largest wood products companies changing their forest management practices and, hopefully, their villainous reputations as well.

During three days of tours in the Pacific Northwest this summer, leaders of several timber companies did not shy away from discussing clearcuts. They said the selective tree harvesting techniques that leave behind one-third to two-thirds of the trees on a given plot are the wave of the future, where affordable.

But they also defended clearcutting and in some cases were adamant about their intention to continue....

"People see a clearcut and, not understanding what is going on, they just view it as a terrible thing. Actually it is like an empty field that has been harvested and plowed under," he said.

"Next year's crop in Kansas comes up the next year. Unfortunately in the Pacific Northwest, with the next crop of Douglas firs, it is twenty years before it starts to look like a forest," Steve Rogel, president of Willamette Industries, said.

But environmentalists say native forests aren't anything like a row of crops. They say a tree farm is a monoculture — the same species of tree throughout — while a native forest is full of biological diversity....

Jack Creighton, president of the Weyerhaeuser Company, said the controversy over clearcutting is about emotions, not science....

"We think on our property we have tremendous biological diversity," said Creighton, whose timberland covers 5.6 million acres in the United States and 2.8 million acres in the Northwest. "But it's an emotional issue. People's feelings run very high," he said. (The Associated Press, December 19, 1994)

The reason for the public's outcry against clearcutting is all too clear: directly after clearcutting a forest often looks like Armageddon. Little wonder that the timber industry, with its ability to seemingly "destroy" a forest, is vulnerable to controversy. No explanation thus far offered has encouraged the environmentally conscious public to wait patiently for the "destroyed" forest to regrow. The public's dislike of clearcutting transcends regions, species, and ownership. An exception, perhaps, is the huge timber plantations in the Southeast part of the country, and increasingly in other areas, where forests are grown solely to be harvested.

It is difficult to defend their practices. With such graphic evidence as a clear cut, the video documentary, *The Forest Wars,* produced by the forest industry, on the debate over management of United States private and public forests can't compete with a persuasive video such as *Logs, Lies, and Videotape,* produced by groups opposed to salvage logging. "The beauty, the tragedy, the hypocrisy, the waste, and the pain" of *The Forest Wars,* as depicted by the American Forest and Paper Association, and other industry organizations, lacks the passion of *Logs, Lies, and Videotape.*

A recent survey found that younger foresters and executives of forest industry companies, who were taught to care about the environment from kindergarten days, view clearcutting as a major factor in the negative image of the industry. However, their older bosses are more likely to fault the media reports that have turned younger foresters against clearcutting instead of examining the environmental impact of the practice. Some foresters disagree with the fisheries biologists that clearcutting is responsible for the Chinook salmon predicament.

Thus, litigation and media coverage of clearcuts support the belief that clearcutting is a major reason for the public's displeasure with the forest industry. Ignoring the environmental reasons that may not concern hikers, tourists, or homeowners, clearcutting has a bad name because a clearcut forest looks as dismal as if hit by a bomb. It transforms peaceful vistas into war zones. A fresh clearcut is an eyesore, except perhaps to the forester who appreciates its silvicultural value.

However, modern log harvesting, contrary to the cut and run of an earlier era with a larger resource base and different needs, is hardly the casual, wanton process the public so readily condemns. A brochure published by MacMillan Bloedel, one of the Canadian timber giants, lists some of the questions studied before logging an area.

Concerns for the environment include:

- Is the area a watershed supplying community water?
- Are there any fish spawning near the area to be logged?
- What wildlife lives in the area? Will they be endangered by logging?
- Are the soils stable, or sensitive to erosion?
- Is this an area of unique beauty?
- Are there any archaeological sites in the area?

The brochure explains MacMillan Bloedel's rationale for clearcutting. If the company practiced selective logging — taking a tree here and there — rather than clearcutting, the consequences to small operations would be a fragile stand of trees left after selective logging and danger to loggers from unstable trees. Moreover, the company claims, clearcutting is essential to regenerate forests of Douglas fir and similar species that need open ground and sunshine; and selective logging costs more than clearcutting.

However, MacMillan Bloedel's rationale, in the public mind, does not compensate for the barren moonscape left by clearcutting. Clearcutting continues as an issue alienating the public from the forest industry.

Emotions boil when a company clearcuts a forest to produce paper. Currently, British Columbia's remaining stands of virgin timber are the target of clearcut opponents.

Robert Kennedy Jr., of the Kennedy dynasty, a lawyer for the Natural Resources Defense Council, and one of the celebrities carrying the torch against clearcutting in British Columbia, rails at the clearcutting of virgin timber from Clayoquot Sound, called "the most pristine example of coastal temperate rainforests."

"Over half of the wood cut in Clayoquot Sound ends up in the United States," Kennedy asserted. "There are whole mountainsides that look like they were the subject of severe carpet bombing. The newsprint users are the most visible users, and they're the easiest to target."

Clayoquot Sound has won worldwide attention for its efforts to create a backlash against visible newsprint users for buying paper from clearcut logging in the Clayoquot. The *New York Times* and the *San Francisco Chronicle,* who have published editorials against clearcutting, were high-profile victims of the backlash. Environmental groups paid for full-page ads in the *New York Times* decrying the paper's use of clearcut paper, presumably printed on pages from the trees they were trying to protect. The *Wall Street Journal,* Knight-Ridder and Gannett papers have also shared the unwelcome spotlight.

The January-February 1994 issue of *Sierra,* the magazine of the Sierra Club, published an article, "Mea Pulpa," on the paper used in their magazine. When the author of "Mea Pulpa" dug into the source of the paper the magazine was printed on, he was shocked to find that 15 percent of the paper pulp came from liquidating British Columbia's "magnificent ancient-forest stands." The article concluded that there are no 100 percent ecologically sound choices for magazines printed on coated paper. *Sierra* reported that it was unable to find any manufacturer to supply recycled paper (with at least 30 percent post-consumer content) that contains totally chlorine-free (TCF) virgin pulp from well-managed second-growth forests or plantations. (Post-consumer waste is the industry jargon for the reuse of office wastepaper, newspapers and other paper products.)

Sierra calculated that, "By using paper containing 10 to 20 percent post-consumer waste for the past three years, *Sierra* has saved 5,800 trees, 2,403,300 gallons of water ... and reduced air pollution by some 20,600 pounds."

Clearcutting, old-growth, rainforests, toxic chlorine, smog-producing chemicals are five environmental "sins" rolled into producing paper. It's a big order for a logger to handle when sweating or freezing in the woods!

Clearcutting decisions remain a painful example of the unsatisfactory options faced by companies when harvesting trees. According to some silviculturists, tree species that require sunlight, the Douglas fir and pines, for example, need to be clearcut for regeneration. Clearcutting also offers foresters the most options for managing the remaining timber in a forest. But there are other equally qualified scientists who reply, "Nonsense!" to the clearcutting claims, insisting that thinning stands of trees, or harvesting selectively, can accomplish the silvicultural missions and leave the forest looking attractive.

Bio-ecologist Dr. Edward F. Hooven conducted an in-depth study of the effects of clearcut logging on wildlife population, concluding that, "Objections to clearcutting of Douglas fir have included possible detrimental effects upon wildlife. Observations show that the ... changes occurring from clearcut logging are temporary and beneficial to forest wildlife." Citing hysterical public objection to clearcut logging with phrases such as "damage to pristine beauty," "rape of natural resources," and "harm to wildlife," Hooven maintained that clearcutting of Douglas fir in Western Oregon is beneficial to wildlife, including fisheries, and is good for replacing forest stands and forest stewardship.

Although Hooven's and other research may be valid,

the public is not accepting the devastating appearance of a clearcut. Clearcutting, visible to motorists along the country's roads and plane passengers from the sky, instantly identifies the timber industry at work. Clearcutting has become the "Remember the Alamo" or "Remember Pearl Harbor" of the environmental community.

The theme of the Sierra Club's handsome coffee-table book, *Battlebook Clearcut — The Deforestation of America,* was cited in a Congressional hearing as covering "one of the most controversial questions ever to involve the nation's federal timberlands." Clearcutting became the focus of litigation against the Forest Service and the flagship of the environmental community's war against the forest industry.

The American Forest and Paper Association (AF&PA), whose 425 forest and paper company members produce 95 percent of the paper and 65 percent of the United States' solid wood production, recognized the public's irritation with the appearance of clearcuts when it developed its Sustainable Forest Initiative (SFI) Guidelines. With a new sensitivity to the visual impacts of clearcuts, Guideline 5 called on AF&PA members to "minimize the visual impact by designing harvests to blend into the terrain, by restricting clearcut size, and/or by ... judicious placement of harvest units to promote diversity in harvest cover." People are not patiently waiting for the AF&PA Guidelines to take effect. The public's exasperation with the forest industry's clearcutting practice has given birth to petitions in both the East and the West to legislatively ban clearcutting once and for all. The environmental advocates railed against "even-aged forests" — woods where trees are all one kind and one age after an area that has been clearcut of its natural trees and vegetation is replanted with tree seedlings. They maintained that clearcutting damaged the ecology. In the face of strident criticism, the forest industry continued to assert that the visual impact of clearcutting was short lived and in a few years would not be distinguishable from a natural forest. The public was not persuaded by the forest industry's explanation that clearcutting was the cheapest way to harvest and that the tree plantations that replaced the clearcut forest provided better growing conditions.

The U.S. Forest Service, meanwhile, under pressure from the public for a greater voice in forestry decisions and the mandate for public participation provided by the National Forest Management Act (NFMA) of 1976, expanded an "Inform and Involve" program to explain its forest practices. The records do not indicate that the participation mollified the public.

Environmental advocates sneer that these clearcut restrictions are merely "beauty strips." Clearcutting is firmly fixed in the public mind as a negative impact. The perception of harm from clearcutting has led to initiative petitions in both the East and West to legislatively ban all clearcutting.

Widespread clearcutting began about 1964; by 1969, 63 percent of the timber cut in the West Coast fir forests was clearcut. A torrent of disputes, innuendos, and scientific papers have flowed around the issue for years. Multiple use was pitted against the advantages of even-aged management; the need for wood was supported by the assertion that the visual impact of clearcutting was short-lived and in a few years would not be identifiable. When faced with a mounting barrage of bitter criticism, the Forest Service determined to solve the problem by expanding its

Information and Education Program to inform the public about clearcutting practices.

Intensive studies with often conflicting conclusions have examined the impact of clearcutting on fire, insects and pests, regeneration, wildlife, watershed management, soil capacity, fish, and other denizens of the forest. In order to gain perspective, it's important to remember that this issue was exacerbated by pressure from several administrations in Washington to increase the timber cut from the national forests.

Professional foresters, the academic community, the Forest Service, and industry trade associations have each attempted to "educate" the public about the need and virtues of clearcutting. "If only the public understood the benefits of clearcutting," the reasoning went, "they would approve the practice." Education doesn't appear to be working. For example, when *Battleground Clearcut* appeared in 1994, illustrated with devastating photos of clearcuts, the timber industry cried "Foul!" pointing out that many of the photos were of old clearcuts which had already regenerated to second-growth forests, and others were incorrectly identified.

Those who assert belligerently that clearcutting is not an issue and that the public will approve the practice if they understand why clearcutting is necessary, ignore the feuds about clearcutting between the forest industry and environmental activists. In fact, so powerful is the negative perception of clearcutting that the courts have sometimes judged clearcutting after little discussion. In 1976, U.S. District Court Judge William Wayne Justice, in the case *Texas Committee on Natural Resources versus Butz* (then U.S. Secretary of Agriculture), concluded its findings of fact after only one day of hearings with these statements about clearcutting:

"Clearcutting results in increased fire hazard, because fire hazard is a function of fuel moisture content."

"Clearcutting impairs the productivity of the land, because it causes accelerated erosion and loss of important topsoil."

"Research indicates that clear-cutting results in accelerated loss of nutrients."

"Clearcutting impairs and reduces the amount of habitat essential to wildlife."

"Clearcutting impairs the productivity of the land by reducing the number of species. Clearcutting and even-age management may result in the hasty liquidation of high-quality timber and, therefore, lead to a future scarcity of high-quality wood."

The scientific merits of these statements would ignite a bonfire of disagreement among forest ecologists and silviculturists. They illustrate the public perception of clearcutting upon which Judge Justice apparently based his findings after only one day of hearings.

CHAINSAWING OLD GROWTH

The mention of old growth and wilderness also raises adrenaline levels in both the environmental community and the forest industry. Many foresters confess that they don't understand the reverence for old growth. Foresters work with the cycle of life — birth, growth, maturity, old age, and death. Old-growth timber provides highly prized, wonderfully clear, strong lumber. When trees are considered "old" depends upon the species. Alder, the Western hardwood until recently considered a pesky weed and now prized for cabinets and furniture, is "old" at fifty years. Thus, after fifty, the wood begins to decay beyond its use as lumber. The majestic Sequoias, Douglas firs, redwoods, pines, oaks, and maples may, with luck, gracefully last hundreds — perhaps thousands — of years.

To the forester who studies how to obtain the best grades and most lumber from a stand of timber, a decayed tree is a liability. To the ecologists, decay is part of the ecology of the forest. Ecologists value old growth, either dead or alive, so long as the death is "natural" and not induced by the chainsaw.

It's clear that old growth contributes to the mystique of the wilderness. Naturalists see wilderness and old-growth groves as a special place where people can seek solace and solitude. Worshipping them is an experience with a religious, spiritual quality; they are the temples and sacred ground of many environmentalists. Wilderness proponents believe that America's culture has been shaped by the untouched forests. Certainly, wilderness and old growth have been the inspiration for American musicians, artists, and writers. Thoreau's pronouncement that "in wildness is the preservation of the world" has become the code of preservationists.

To other Americans, brought up in the human-centered tradition of conquering the wilderness, the notion that man may be a visitor but should not leave his footprints on any place of earth is foreign. William Tucker, a writer and social pragmatist, asserts that wilderness areas are elitist preserves, a "misguided and romantic ecological ethic." To the forest industry, old growth and wilderness set-asides are part of the attrition of the forest resource upon which the industry is based. The arguments emanating from both camps are likely to continue to be linked with clearcutting and salvage logging with its infamous "rider."

LITIGATION OPPORTUNITIES

Since the advent of the environmental movement, litigation has been a powerful tool in achieving environmental goals. When lobbying fails, when a campaign to win the public to its proposals fails, both environmental organizations and the forest industry resort to litigation. The federal government has been the principal defendant in litigation on environmental laws.

The Environmental Impact Statement requirement of NEPA legitimized lawsuits against the Secretary of Agriculture, formerly Earl Butz, acting as a surrogate for the United States Forest Service. The suits primarily disputed the increased timber cutting on the national forests. About a thousand suits were filed in the early 1970s against the U.S. Forest Service, as many as seventy cases at any one time, following the passage of NEPA. The Organic Act of 1897, passed by Congress, remains the fundamental legislation for the management of the nation's publicly owned forests. As a result of this trend in litigation, many forest management decisions were made by the courts, not professional foresters.

Moccasin Basin in the Teton National Forest, Monongahela, Bitterroot National Forest, Boundary Waters Canoe Area, the Tongass National Forest in Alaska, and the Organic Act of 1897 became household words, although most of the population did not know where the forests were located and had never heard of the Organic Act. The Bureau of Land Management (BLM) Organic Act of 1976, officially the Federal Land Policy and Management Act, added strength to the ownership of federal land by the people of the United States.

The Western national forests, containing the major uncut softwood stands in the nation, became the target of litigation, and also received the bulk of media coverage. The constant barrage of attention to the federal forests has led many people to the unwarranted conclusion that all the worthwhile forest land and timber is located in the federal forests and under the jurisdiction of the Forest Service. Forestry professionals fear that this erroneous conclusion will be a major barrier to providing the wood the country needs.

Currently a main issue in the courts revolves around the environmental community's vigorous objections to the "Timber Salvage" bill, which captured public attention in the summer of 1995 and 1996. The salvage logging issue underlines the lack of trust of the environmental community in the forest industry.

On July 27, 1995, President Clinton signed a budget bill that slashed $16.3 billion from the federal budget. The bill might have vanished into obscurity had not an innocent-sounding "rider" been attached to it to release timber from federal lands that previous environmental legal challenges held up to protect old growth, the spotted owl and the marbled murrelet. The rider was described as a health measure, removing dead, dying, or diseased trees and leaving behind healthy trees — surely a win-win situation if there ever was one in a forest controversy. To speed recovery of the wood before it deteriorated further from fire and disease, an emergency period was proposed. During that time the government was given the latitude to proceed with timber sales without the customary required reviews.

Administrative appeals were not allowed, and courts would be denied the power to issue injunctions against sales. By also suspending existing environmental laws during the emergency period, environmental groups could not litigate against any salvage logging. The Forest Service spent most of the previous winter preparing required environmental impact reports on the sales of timber damaged many months before. Timing is a critical factor, as major species like Ponderosa pines "blue" after dying and quickly lose value for producing lumber.

Environmental conservation organizations vigorously opposed the salvage rider, accusing the administration of providing a quick source of lumber and jobs to the timber mills and the "powerful" home-building industry. Quoting from newspaper accounts of the infighting over the Salvage bill, environmental leaders stated their fears about the consequences of the bill. They feared that:

- The government would expand *harvesting into roadless areas.*
- The harvest could *endanger fish streams.*
- It gave federal forest managers too much discretion.
- It freed the Forest Service and the Bureau of Land Management from the responsibility to comply with environmental laws.
- The definition of salvage was so broad that it opened the door to logging in the remaining *old-growth forests.*
- It did not require *public input* into the decisions.
- It banned appeals or *judicial review* on these and future sales "opening the door too wide for imprudent actions too hastily considered."

- It allowed completion of previously sold timber sales held up but not awarded to protect critical nesting areas of *endangered sea birds* called marbled murrelets.

On the other side of the issue, the log-hungry timber industry, frustrated after having its hands tied in harvesting the timber it had already bought from the Forest Service, saw the rider as a necessary, sensible solution. Meanwhile, the environmental community, deprived of the critical ability to litigate, saw the rider as a call to action. Ultimately, the salvage rider renewed the simmering forest wars. The May 2, 1995, editorial in *The Oregonian,* Oregon's major newspaper, called for Congress to "incorporate legally binding requirements for the protection of other critical values, such as salmon, endangered species, and critical ingredients of a healthy forest."

Protests, civil disobedience, and negative press followed logging activities on Western forest lands, with the timber industry again emerging as villain. The backlash was stinging. After the intense national criticism from the environmental community, one year later more stringent guidelines for salvage logging were enacted. During this controversy, the Eastern hardwood forests were innocent victims in the web of the backlash. Salvage operations applied to eradicate the Gypsy moth in Eastern forests, where old growth is practically non-existent, have not prompted the same resistance, but they still must deal with the backlash from the West.

ENDANGERED SPECIES

Robert T. Lackey, deputy director of EPA's Environmental Research Laboratory, describes how the *anthropocentric* view of the earth, or a belief that all decisions affecting ecological systems benefit humans in some way or other, underlies the government's right to regulate logging and sacrifice to save endangered species:

> To be sure, we may preserve wilderness that few visit, protect from extinction obscure species that have no demonstrated utility, and spend vast sums to restore habitats for species of limited economic value. All these efforts provide benefits to people; the benefits may be non-economic and non-monetary, and may be only to buy some indeterminant form of future insurance, but they all benefit humans.... The entire regulatory framework to protect ecosystems is set up under this assumption. (Robert T. Lackey, September 1994)

Professor Lackey's explanation of the rationale behind the Endangered Species Act (ESA) may appease intellects, but to the loggers whose chainsaws must lie idle, and sawmill workers who have no logs to saw or lumber to pull on the green chain, the ESA is an abomination that sacrifices their livelihood for some unknown future. And it's important to remember that the ESA affects both publicly and privately owned forests.

A 1995 poll found that 45 percent of Americans believe the Endangered Species Act (ESA) has gone too far, and 41 percent are satisfied with the Act as it is and want no changes. Interesting to note is that 65 percent of those polled believed that cost should be considered in saving species.

It appears that a great many Americans share the views of Professor Lackey, but, as he also points out, a smaller but powerful group of ecocentered, earth-centered people believe that humans are only one species and are no more important than others. All species must be treated equally. This "deep ecology" view, shared by some religious and philosophical creeds, plays a pivotal role in politics. The ecocentered view believes that the environment should be brought into the legal system as a holder of legal rights. In the debate about ecological risk assessment they forcibly argue that such assessments are a form of ecological triage.

The Endangered Species Act (ESA), signed into law by President Richard Nixon in 1973, introduced a new category of conflicts between human interests in using forests and ecological interests: stamping out forest fires threatened the rare Karner Blue butterfly; harvesting timber threatened spotted owls and marbled murrelets; hunting jeopardized the whooping crane. The list of endangered species is constantly growing with new discoveries.

The ESA directed the Fish and Wildlife Service, a branch of the Department of the Interior, to maintain a list of species that are either endangered (in imminent peril of becoming extinct) or threatened (likely to become endangered in the near future). The list was not an innocuous piece of paper, as so many lists are. Once a species is listed, people are not allowed to *"harass, harm, pursue, hunt, shoot, wound, kill, trap, capture, or collect"* members.

Bob Beschta was standing on a gravel bar in the Lamra River along Yellowstone National Park's hauntingly beautiful northern range.... But Beschta was distinctly not enjoying the scenery.

"This is just off the charts," he said, shaking his head as he stared at a spot where the meandering river had cut deeply into the black silt on either side. "The banks are not capable of holding this channel in place anymore." Beschta, a hydrologist at Oregon State University's College of Forestry, had just become the latest convert to a growing cadre of ecologists, river experts, foresters and rangeland scientists who are convinced that there is something seriously wrong in the nation's oldest park.

For years, environmental groups have fought a series of bitter actions against forces they see besieging the park.... In their view, mining, logging, ranching and hunting are the enemies that have placed this sacred ground at peril....

But to Beschta and many of his fellow scientists, the real danger comes from a very different source — one squarely within the park itself. They paint a bleak picture of an ecosystem literally unraveling, as stream banks erode, woody shrubs disappear, stands of aspens and willows die, and many once-abundant species from beaver to birds dwindle. The culprit they point to is elk; more particularly, National Park Service policy that allowed elk numbers to increase from 3,100 in 1968 to some 20,000 today on the northern range.

"What they're changing this into is a lawn. There's more to an ecosystem than grass," says Charles Kay, a wildlife biologist at Utah State University and a leading critic of Yellowstone's management.

Kay has collected reams of evidence that he says documents the park's slow — and in some cases not so slow — ecological death.... (*U.S. News and World Report,* September 16, 1996)

Another viewpoint comes from Richard Starnes, editor of *Outdoor Life* magazine, who argues that federal regulations are not the answer. He believes that professionals in the state and national agencies should be responsible for determining and carrying out the policies. ("The Sham of Endangered Species," *Outdoor Life,* August 1980.)

The endangered species need not have any value to humans to be cherished and protected, but should qualify for protection simply because they exist and have existed for a long time, according to the principle designated the "Noah Principle," by David Ehrenfeld, an ecologist at Rutgers University. Interest in endangered species was originally inspired by Robert Porter Allen, an expert on whooping cranes, often called the "poster child" for biodiversity, or the symbol of everything natural that might be lost. Allen was able to create a tempest over the potential disappearance of the whooping crane, initiating the first federal programs to protect endangered species in 1966, and later the Endangered Species Act itself in 1973.

The politics involved in the passage of the endangered species legislation included federal wildlife biologists, four presidents, the Fish and Wildlife Service, the Department of the Interior, prominent Republican leaders, several influential senators, assorted prominent ecologists, and the International Union for the Conservation of Nature and Natural Resources (IUCN). After it was passed, political maneuvering continued as conflicts between the preservation of endangered species and human welfare accelerated.

The original 1964 endangered species list included sixty-two animals, thirty-six birds, fifteen mammals, six fish, and five reptiles and amphibians. Since the passage of the Endangered Species Act, 721 species have been added to the list. Another component of the legislation was that the U.S. Supreme Court upheld federal rules forbidding destruction of habitat vital to an endangered species, even on private property.

The decision that protecting endangered species is the law of the land created a new series of scientific controversies, with credible experts taking both sides. Lewis Regenstein, vice-president of the Fund for Animals in Atlanta, Georgia, argues that the Endangered Species Act needs more stringent federal regulations, claiming that the ESA is actually exploiting rather than protecting wildlife, and threatens to hasten the disappearance of endangered species. ("Endangered Species and Human Survival," *USA Today*, September 1984)

The Endangered Species Act had more clout than ever imagined by the originators and Congress. At times it appears to be a blank check to obstruct any construction. A highway in Oklahoma was detoured to save the habitat of an obscure beetle called the American burying beetle. Logging was prohibited in large sections of national forests to preserve the habitats of spotted owls, and assorted other birds. The United States Supreme Court determined in the notorious *Tellico Dam versus the Endangered Snail Darter* decision that Congress decreed the value of endangered species as "incalculable; therefore, one species could not be considered more valuable than another."

To straighten out the impasses caused by this decision, in 1978 Congress created the "God Committee," an endangered species committee that had the power to permit the extinction of a species when the benefits of a project "clearly outweigh" the benefits of saving the species.

As the repercussions from the Endangered Species Act spread across the nation, many people wondered if Congress knew exactly what the Act meant. Indeed, according to Paul Lenzini, then chief counsel for the trade association of the state fish and wildlife agencies, quoted in Charles Mann and Mark Plummer's book on the endangered species act, *Noah's Choice:* "Few members of

Congress had the foggiest idea of what they were doing. There was no idea that their ox was being gored, so they all voted for it." When the final bill was presented to each Chamber, not a single Senator voted against it and only four members of the House of Representatives cast ballots against it.

Lynn Greenwalt, then-director of the Fish and Wildlife Service, remarked: "They [Congress] weren't thinking about dung beetles. They were thinking of huge grizzly bears and bald eagles and stately monarchs of the air." The law gave the Interior Department almost total powers over federal land and considerable control over the 70 percent of the land not owned by the government. Thousands of plant and animal species made the suspect endangered species list. Determining which species should be called endangered required months of investigation and documentation. Congress appropriated $41 million for the first five years of the Act.

The jury is still out on the value of the ESA. The stories of the frustrations of builders, ranchers, forest landowners, and homeowners confronted with the fallout of the Endangered Species Act will fill many volumes. In some cases it has had a reverse effect in protecting species. For example, the Balcones Canyonlands National Wildlife Refuge near Austin, Texas, is surrounded by dozens of private ranches. To discourage the endangered small black bird vireo from moving to their property, ranchers near the reserve are reported to be deliberately mismanaging their land to avoid providing habitat for the birds. *Noah's Choice* quoted Deborah Holle, a wildlife biologist at the refuge: "If you do something to attract those birds to your property and then want to raise goats or sell your property,

Fish and Wildlife [agencies] will say you can't do anything now. The way it is, it would be kind of silly to help the birds out."

If gauged by how many endangered species the Act has saved, it appears to have possibly improved the prospects for 10 percent of the species listed as endangered. However, many observers believe that the Act has accomplished little for biodiversity and are proposing revisions. Still, one conclusion is irrefutable: The Endangered Species Act has endowed biologists with the kind of power that political scientists would envy.

SAVING THE RAINFORESTS

Another important debate has arisen because the potential extinction of rainforests has captured the attention of the world. Few people outside biological circles previously heard about rainforests, especially tropical rainforests, before the early 1980s. When Terry L. Erwin, entomologist at the National Museum of Natural History, warned that "We are rapidly acquiring a new picture of Earth. Millions upon millions of nature's species [are] on the verge of being replaced by billions upon billions of hungry people, asphalt, brick, glass, and useless eroded red clay, baked by a harsh tropical sun," he painted a grim version of the environment — the tropical rainforest. And then another dimension to the issue was added when saving the tropical rainforest became fashionable.

These days we can save the rainforest by buying rainforest candy, rainforest ice cream, and rainforest cereal. For example, a cereal commercial shows a "politically correct" family beginning the day by buying the cereal because 8 percent of the ingredients are provided by the indigenous peoples of the Amazon. (The rationale is that the natives can use the money they make from the cereal to stop cutting their trees.) Rainforest flower seeds are available in garden stores, and the actress/singer Madonna adds her voice by singing "Don't Bungle the Jungle" at rock benefits.

According to many scientists and writers, the survival of the rainforest became the key to saving the earth's species. A University of California ecologist, Jared M. Diamond, writes that many rainforest species cannot exist outside the rainforest; thus, cutting the rainforest would lead to global extinctions. Assorted scientific reports transformed the tropical rainforest from a biological concern to a personal nightmare. Paul and Anne Ehrlich published *Extinction,* warning that the exponential increase in the world's population would practically deforest the world by the year 2025. The Ehrlichs warned that humans are "no more immune to the effects of habitat destruction than are the chimpanzee, Bengal tiger, bald eagle, snail darter, or golden gladiolus." Extinction predictions, backed by "scientific" calculations, came tumbling out in learned papers.

It didn't take long for the terrifying scenarios to hit home. European countries banned the import of the exotic wood from the Amazon tropical forests, and the Rainforest Alliance was formed. The Woodworkers Alliance for Rainforest Protection (WARP) emerged in 1989, focusing on the tropical rainforest, "which was being destroyed at an alarming rate." A number of large charitable foundations sprang up, funding projects relating to the rainforest.

Sustainable forest management, certification by an objective organization — a third party — that a forest is managed sustainably, and a quickened interest in sustaining forests resulted.

Then in the 1990s, interest grew when the rainforest assumed a new role as a source of newly discovered therapeutic plants as researchers sought cures for cancer and AIDS. Gordon Cragg, director of the Developmental Therapeutics Program, remarked that cutting down the rainforest "would have a significant effect on our chances of discovering new agents."

The forest industry has been relatively silent about its role in the rainforest, except for pointing out that logging is responsible for less than 5 percent of the deforestation of the tropical rainforest. Few people are standing up and declaring that the rainforests should be cut. "Preserve the rainforests" has become enshrined in American slogans along with "Remember the *Maine*" and other historic remembrances calculated to stir the emotions.

SAVING JOBS OR BIRDS

Alan Thein During, of the WorldWatch Institute, summarizes the saving jobs versus saving the environment issue in his book, *Saving the Forests — What Will it Take?*: "The main argument that entrenched interests will make against sustaining forests is that if the trees stand, jobs will be lost. In the Northwest, the question of whether to protect the ecological integrity of old-growth forests was popularized as a conflict between jobs and the endangered northern spotted owl." During accuses the timber companies of fighting, not for jobs, but for profits, and justifies his accusation with the statement that: "The companies had been quietly trimming payrolls for ages as advancing technologies made labor-saving possible." He concludes that: "Their jobs are no more a reason to continue deforestation than jobs in weapons plants are a reason to go to war."

The loggers and sawmill workers whose jobs have been sacrificed don't see it that way. They maintain that the closure of the sawmills because they can't obtain logs deprives them of their ability to earn a living and support their families. They resent the people they call "enviro elites," who are forcing them into a new way of life. They bristle at the notion that saving some birds and trees are more important than saving their lives. Having lived intimately with nature in the woods, they feel that they are the real environmentalists.

The federal and state governments are pouring money into salvaging the rural communities that have depended on the forests to sustain their income, and retraining the loggers and millworkers whose jobs have disappeared to work in the tourist industry, or factories that relocate to their areas. The forest products industry, meanwhile, is undergoing a massive restructuring to continue to produce wood products for the nation's needs and to find the money to build the expensive new plants to produce the reinvented wood described in a later chapter.

THOSE WICKED PESTICIDES AND HERBICIDES

Paul Muller's discovery that DDT could eradicate insects won him a Nobel Prize and fueled agriculture's Green Revolution's promise that we could now grow enough food to feed the world and prevent mass starvation. Later discoveries of the insecticide malathion added to the arsenal of chemicals that aided humans in conquering nature. But by the 1950s pesticides were acquiring a reputation as carcinogens, setting the scene for the Delaney Amendment, banning any trace of cancer-causing pesticides in processed foods. Rachel Carson's *Silent Spring*, linking

DDT and similar pesticides to widespread damage to the environment and human health, branded herbicides and pesticides as dangerous substances.

Fear of toxic pesticides led to alternative pest control methods that avoided toxic pesticides. Integrated Pest Management (IPM) uses physical barriers, biological controls, and non-toxic chemicals with synthetic pesticides as a last resort when nothing else has worked. With missionary zeal IPM advocates organize pesticide education. Typical is the San Francisco Pesticide Education Committee, whose president works to correct the notion held by many people that "products from the chemical industry are the only way to live modern lives."

Despite the fears about the dangers of pesticides, the reality is that damage from insects and disease outstrips damage caused by wildfire. As much as 60 percent of some forest land is killed by insects and disease. After many years of touting Smokey the Bear fire prevention and suppression, foresters are changing their tune. Now controlled fire has been graduated to a forest health measure.

Experts on both sides of the issue hotly debate the use of pesticides and herbicides. The chief of the United Nations Food and Agricultural Organization warns that pesticides are the key to the massive agricultural development needed to prevent starvation in the world. On the other hand, a well-known ecologist contends that dependence on pesticides has often worsened pest infestation and creates new health problems.

THE BELEAGUERED U.S. FOREST SERVICE — BETWEEN A ROCK AND A HARD PLACE

A discussion on controversies in the forest would be incomplete without the saga of the U.S. Forest Service.

From a reputation as the top land management agency in the world to being denounced as "an agency gone sour," the role of the U.S. Forest Service has bounced around like an inflated balloon. Everyone seems to be mad at the Forest Service: the forest industry, environmental groups, forest-based communities, and the confused public.

The Forest Service was born out of controversy, but for a golden forty-five years it enjoyed a benign neglect, free to use its professional knowledge of forestry to protect forests from fire and provide professional silviculture as stewards of the national forest lands. Its leadership in management of national forests brought professionalism to all public and private forests.

The concerns and accusations that led to the creation of the Forest Service have a familiar ring today: Between 1871 and 1897, individuals and companies were so hungry for land for farming and timber that Congress considered 200 bills to protect the forests. The bills aimed to reserve forests to protect the headwaters of navigable rivers, block public access to forest reserves, regulate timber sales on public lands, protect water for irrigation, and protect forests from fire.

Most of the bills died before Congress could vote, but two finally made it through the maze: The Forest Reserve Act (1891), and the Organic Administration Act of 1897 (the Forest Management Act), directing management of the forest reserves. A resident of the logging town of Darrington, Washington, wrote to the Arlington newspaper: "I believe that the reserve was created for the benefit of a few lily-fingered gentlemen who want a place to hunt and fish. I can imagine no other reason." A great-grandchild might be writing the same letter today in response to some of the current actions by Congress.

In the 1800s the practice of replanting trees was rare; and bills to encourage tree planting and preserving forests on public land engendered storms of ugly debate but failed to pass. Although tree planting is routine now, the debate on clearing forests still continues.

To prevent land fraud and job losses, the 1897 Forest Management Act created the General Land Office (GLO) in the Department of Interior to manage the reserves.

The common practice of stealing public timber by railroad companies needing crossties, added to the losses from forest fires that destroyed more timber than was stolen, angered Gifford Pinchot, from the Department of Agriculture, Forestry Division. He exclaimed after a visit to Montana, "Out here a man is hung for stealing a horse, but a corporation stealing millions of board feet of timber is looked upon with tolerance."

The first predicted "timber famine" turned into a rallying point for legislation, with then Interior Secretary Carl Schurz warning in 1875 that another twenty-five years of logging would leave the U.S. "as completely stripped of her forest as Asia Minor today."

Public forests, especially those in the West, became a political football, with bills quickly passed to accomplish some mission, often encouraging fraud and abuse with results Congress had not anticipated. Even the Timber Culture Act of 1873, rewarding anyone planting a certain number of trees, backfired. Subsequent repeal of the Act appears to be a forerunner of current Congressional action. Twenty-three amendments, add-ons, and other items were tagged onto the repeal bill. Indiana Representative Holman sneaked in a twenty-fourth amendment at the last minute, a rider that authorized the president to set apart "public lands covered in whole or part with timber or undergrowth, whether commercial value or not, as public reservations." Congress passed the Act without realizing that it gave the president far-reaching power to remove forest land from the public domain.

The president needed professional assistance to manage these lands, assistance not available from the political appointees in the Department of Interior. President Theodore Roosevelt, a Republican and nature enthusiast, used his political muscle to transfer the forest reserves out of the Interior and created the new Forest Service in Agriculture with Gifford Pinchot as chief. The new agency had the authority for regulation, enforcement, and budget, and the means to achieve the vision of sustained yield harvest rather than "cut and run." The plan was for the U.S. Forest Service to plant trees and nurture them into healthy forests. By the 1920s, private lands followed suit.

In the 1920s, the timber industry urged the Forest Service to curtail logging so as not to flood the market with timber. At the time the Forest Service logged only in localities where communities depended on national forests. It protected forests from fire, and purchased cut-over lands in the South and East for the purpose of reforestation. The Forest Service became known as a model agency, earning trust and respect. Of course, like most bureaucracies, politics ruled behind the scenes, with dramatic late-night additions to the forest reserves. Westerners, outraged by "Pinchot's midnight land grab" adding nineteen million acres of the finest timberland in the West to the national forest, promised dire revenge.

William H. Taft's ascendancy to the presidency changed control, and Pinchot was fired, a victim of disagreements with the president. However, Pinchot's legacy of sustained-yield forestry remained, as did his influence. The second of the "timber famines" forecast in the United States came from Pinchot, who warned that the famine would strike unless modern forestry was adopted. The former President Roosevelt supported the prediction: "The United States has already crossed the verge of a timber famine so severe that its blighting effects will be felt in every household in the land," Roosevelt proclaimed.

In 1939, the third prediction of a timber famine came from Chief of the Forest Service, William B. Greeley. Despite political machinations, things went smoothly in the woods until World War II. The war effort consumed more tons of wood than steel. In fact, by the end of the war, private timber holdings were severely reduced, and the remaining good wood was on federal lands. At the same time, society started changing. The Baby Boom was born, and Americans were desperate for homes. Meanwhile, working conditions were changing, and Americans also wanted scenic beauty for vacations.

The Forest Service was thrown into the lion's den, trapped between the jaws of economic necessity and environmental values. Congress tried to solve the dilemma with the Multiple-Use Sustained Yield Act of 1960. Its purpose was to balance "recreation, range, timber, watershed, and wildlife and fish purposes."

As different forces faced the agency, the forestry school graduates with military experience on the U.S. Forest Service staff knew how to grow and cut trees but were not trained in public relations. Thus, they were not prepared to respond to the new public ideas about the forest. The requirements of the new acts in the early 1970s, the

National Environmental Policy Act, the Resource Planning Act of 1974, and the Monongahela Forest lawsuit began to sap their time and energy. In addition, funds that might have been spent on forest management went to the public participation process and non-timber management. The National Forest Management Act (NFMA) of 1976, providing for an open public process for long-term forest management plans, was dumped on the shoulders of an already burdened staff. Decisions needed the wisdom of Solomon, and Solomon was absent from the ranks. Conceived as a way to reduce conflict, the NFMA failed to reduce conflict; in fact, it provided a forum for it.

The NFMA opened the gates to an invasion of "ologists," specialists who often disagreed over ways to "provide for diversity." Interest groups joined the debate, advocating for hunters or endangered species or aesthetic values or even, at times, logging families. An accessible judicial system confounded attempts to determine outcomes on their individual merits with a series of rulings that gave standing to trees and wild creatures and discouraged extractive efforts. Many lawsuits were filed against the Forest Service, with the green side often winning by procedural error.

The NFMA was based on the theory that good process would eventually lead to good results and agreement. Detailed procedures to achieve the strict compliance that, hopefully, would avoid litigation, took too much time, paper, and money. The results from the Act were that the Forest Service lost its trust and its discretion in implementing changes to the status quo, and never generated the clear direction and support it needed. It reacted by writing more and more ecologically based regulations, which, in turn, drove state regulation of private lands.

The fourth timber famine threat came after World War II as part of the argument for regulating private lands. Attorneys replaced foresters as key people. For example, one Northwest forest company reported preparing extensive Environmental Impact Statements (EIS) on an area to be logged, on the road leading to the land, on the stream below the road and the land, and on the cumulative effects of the road and clearcut on the stream "in the foreseeable future." But since the Forest Service had not prepared a more exhaustive analysis of the additional negative environmental impacts of logging on adjacent private lands to the stream, a court granted an injunction against the logging company citing noncompliance with procedures.

Other experts believe good science is sufficient to drive forest management, and that sophisticated planning tools and staff biologists have the technical ability to successfully balance timber harvest with fragile natural areas. No one has yet determined the unintended consequences of applying technical approaches to problems with values.

For example; a Southwestern forest owner wanted to implement "ecosystem management" while harvesting timber under its forest plan. That meant some islands of habitat (supporting two endangered birds, in this case) might be lost to timber production as lands were reallocated to create a much larger sanctuary elsewhere. Environmental interests threatened to sue under the Endangered Species Act. They didn't trust the Forest Service to protect the threatened habitat because timber interests had historically driven choices and continued to pressure for logs.

The ultimate outcome of the Multiple Use Act was that it inspired more bitterness than goodwill in the forest, confirming a prediction made by Michael McCloskey, then Executive Director of the Sierra Club. The forest industry was still trying to cope with the Multiple Use Act when the newly powerful recreation interests pushed through the Wilderness Act. As the influence of the recreation interests increased, the influence of the Forest Service decreased. In the five years between 1960 and 1965, public recreation use of the Forest Service facilities increased more than 50 percent. Additional national parks were created, removing more timber from multiple use.

Meanwhile, the courts had opened a Pandora's Box by allowing factors other than economic interest as the basis for bringing a suit before the courts. Preservationists were given the green light to stop developments based on scenic or historical or other intangible significance.

Among the courts, new regulations regularly enacted by a busy Congress, active environmental organizations, the environmental consciousness-raising of Americans, and timber-hungry sawmills, the Forest Service was locked in a cage with a dozen lions tearing it apart. The Service reacted by leaning in the direction of its timber supply mandate, continuing the clearcutting practice, even though the public outcry against clearcuts was getting louder. The Agency defended the practice with scientific data but couldn't defend the ugly scars it left on the land.

A trio of hunters temporarily settled the fate of clearcutting on national forests. The hunters claimed that clearcutting in West Virginia's Monongahela National Forest was driving game animals out of the forest. When the Forest Service failed to respond, the hunters, supported by the Audubon Society and the Izaak Walton League, sued the Forest Service in May 1973. The court ruling outlawed clearcutting in national forests. The Monongahela ruling paralyzed timber sales until Congress enacted the Resources Planning Act (RPA) in 1974, and the National Forest Management Act (NFMA) in 1976, that specifically allowed clearcutting in appropriate places.

The elaborate planning process prescribed by the RPA proved to be another case of unforeseen consequences. The diverse science specialties that were required on the planning team changed the makeup of the Forest Service. In 1958, trained foresters composed 90 percent of the force; by 1983 it was 50 percent, and it has continued to drop. As the staff makeup changed, biologists in fish, wildlife, birds, and other natural subjects, sociologists, and archaeologists participated in the foresters' decisions. Then the National Forest Roadless Area Review and Evaluations (RARE II) sent the agency into a tizzy.

RARE II grew out of the Wilderness Act of 1964 that provided a review of previously designated "primitive areas" for suitability or nonsuitability for preservation as wilderness. The intent was to settle the entire matter by 1974, but that changed upon the insistence of environmental advocates that all roadless areas in the national forests be evaluated for setting aside as wilderness. The Forest Service evaluated 56,174 acres in 1,449 areas using a complicated process called RARE and recommended that 274 areas totalling 12.3 million acres should become wilderness. The Sierra Club disputed the adequacy of the Environmental Impact Statement on the wilderness areas and sued the Forest Service. The Sierra Club won the suit, and the process started again as RARE II.

While the RARE I and RARE II studies were conducted, no timber sales could be offered. Every recommendation by the Forest Service that any acreage be released has been litigated. Though it has not merited the attention of the first two RAREs, the Eastern Wilderness Act, passed by Congress in 1975, opened additional areas for wilderness evaluation and might be called RARE III.

An unintended, but perhaps predictable, outcome of the NFMA is revealed in a 1990 survey of Forest Service employees: two-thirds of employees oppose increasing timber output, while over 80 percent support increased wildlife/fisheries habitat improvement projects.

WASHINGTON — Environmentalists thought they'd died and gone to heaven when President Clinton named Jack Ward Thomas to head the Forest Service nearly three years ago.

Thomas, a no-nonsense straight shooter from Eastern Oregon, was the first wildlife biologist named to the post, and many considered his appointment a clear signal that the agency's top priority had shifted from timber to wildlife protection....

But now, as Clinton's first term draws to a close, Thomas finds himself in the hot seat as point man for a meandering logging policy that is every bit as controversial as its predecessors.

While national-forest logging under Clinton is only one-third the level of peak 1980s harvests found to violate environmental laws, the drumfire of complaints from conservationists has not let up. Scores of arrests have been made this summer at logging protest sites.

At the same time, a near-record wildfire season is raging, fueling timber-industry complaints that the administration has not allowed enough salvage logging to clear out dead and dying trees that serve as tinder for such blazes.

"I observe the situation to be volatile," Thomas said of complaints from both sides that Clinton's forest policy is a moving target.... (The Associated Press, September 2, 1996)

Increased fuel loads from a wet winter and the relatively small acreage burned last year combine with insect damage to heat up Oregon wildfires.
More than 47,000 acres has burned this year in Oregon, making 1996 the worst wildfire season in the past twenty-five years.
And with the fire season only half over, state and federal officials say things could get worse. *(The Oregonian)*

As the old-line foresters reach retirement age and the new younger staff who were influenced by the changes in the 1960s and 1970s take over, the Forest Service is likely to be increasingly embroiled in conflicts. While the Service has learned many strategies that don't work, it is not in a position to define a new mission itself. Until Congress refines the mission it wants for the Forest Service, chaos and conflict are likely to continue. Since the industry is prohibited from harvesting on more national forests, the forest industry itself will play a lesser role in the conflicts.

CONFLICT RESOLUTION

Diane Haskell is a partner in a small woodworking plant that produces unique bookends. A long-time resident of her small town, she participates actively in community affairs. When the Forest Service presented a watershed plan for her area, the town split into six factions. Ranchers objected to deer and elk nibbling the winter haystacks saved for their livestock; conservationists in town worried about the impact of the plan on the forest. Community leaders wanted a guaranteed stable water supply; fishermen were concerned about damaging the streams. The local motel owner wanted assurance that fishing wouldn't be diminished as a tourist attraction for his motel guests. With such diverse interests, the townspeople became hostile, and some refused to meet with the others.

Diane dutifully attended the planning meetings held by the Forest Service. Meeting after meeting broke up with no resolution. "At least we talked to each other," said the chairman. Diane snapped, "I'm so tired of talking, talking. We've been talking for a year and haven't accomplished a darn thing!"

"At least we talked!" marks the conclusion of many meetings on forest conflicts. Conscientious citizens are attending meetings, putting green, yellow and red dots next to statements, seeking strategies for promoting sustainable forestry, grazing livestock, changes in logging, or funding a new sewer system.

In forest conflicts involving old growth, sustainable forestry, fishing and recreation, clarification of the issues and a better understanding of the dynamics may help resolve the conflict. In the case of Diane's town, if trust had been established among the parties, conflicts may have been talked out. Demonstration of self-interest of all the parties involved may help bring diverse opinions together.

Those are the relatively easy conflicts to resolve. The path gets rocky when the differences are in the values held by parties. Bookstores are bulging with volumes on achieving resolution or consensus when values differ. In such situations consensus can be, and often is, manipulated by managing and maneuvering to arrive at certain decisions.

When manipulation is successful, the tendency is to scoff, "Politics!" When it's not successful, it has the advantage of usually not being recognized. Social choice or decisions unifying the preferences of individual members into a decision of the whole group uses defined strategies.

MANIPULATION STRATEGIES

Though some people enjoy engaging in conflict, from an efficiency standpoint conflicts consume time and dollars. Conflicts can gnaw at organizational energy like a hungry dog with a bone. Generally, organizations would prefer to reduce conflict, if the option is available. Conflict resolution has grown into a fine art, and organizations and experts have no problem finding conflicts to mediate. Strategies for resolving conflicts involve helping parties understand the facts surrounding the conflict, clarifying the conflict, and demonstrating the self-interest of proposed solutions.

Of the almost fifty conflicts around the forest industry, most involve values and beliefs which are not susceptible to change by learning the facts. Far from being theoretical or philosophical, these conflicts affect people's jobs and community income. Sometimes the issues are whether to install factories or wastewater plants, or affect the availability of foods and products used for everyday living. At some point, the conflicts must be resolved to allow for action or a remedy. Talking may be the prelude, but action — or deliberate inaction — must be the outcome.

When people have fundamental disagreements, the point of view that prevails belongs to the group or individual with the most power to influence the decision. Some elected officials are given the power to make decisions for the group, but usually decision making is in the hands of a committee or a legislative body. Individuals often influence the outcomes by manipulating the group or other individuals.

Manipulation has acquired an undeservedly bad image, as if all manipulation is for unfair or fraudulent purposes. It's important to understand that manipulation is morally neutral; its success depends on the skill with which manipulative techniques are used. Great moments in history are often the products of the manipulation by leaders. Abraham Lincoln manipulated Stephen Douglas in the famous debates that preceded the Civil War. Lincoln asked a question on the relation of slavery to state constitutions that manipulated Douglas into an answer that helped divide his party.

The Seventeenth Amendment on the election of senators passed as a result of manipulating the agenda. Different agendas can produce victories for particular alternatives. The Federal Convention of 1787 that produced the Constitution of the United States provides a model of the use of manipulative techniques to produce majority votes from an extremely disparate group. Potential losses were manipulated into wins by formation and reformation of alliances and redefining the situation to win the support of formerly unsympathetic participants.

The Constitutional Convention had all the dramatic elements of conflict with which the forest industry wrestles today: elitists versus backcountry people, geographical divisions, philosophical differences, varying backgrounds and cultures. The convention succeeded in drafting a remarkable Constitution, thanks to the manipulative skills and strategies of its members.

In the conflicts between the environmental advocates and the forest industry, the environmental community has demonstrated superior manipulative skill and learned how to use these strategies to push its agenda. "Those environmentalists are playing politics," the industry complains. But the truth is that they would be naive not to use their manipulative political skills to achieve their objectives. The industry complaint may be sour grapes because the environmental organizations are better at it than the industry. And, of course, politics is how decisions are made in a democracy.

Chapter 4

Image Repair

NIGHTMARE IMAGES

Sven Justen logs in northern Minnesota. "You can count on Sven," his mill customers say. "He knows what he's doing. You can trust him." Sven cherishes his work record. "I'm leaving my son a good name," he says proudly.

Sven's eighth-grade son tells his classmates that his father drives a truck. He's ashamed to admit that his father is one of those loggers who destroy the planet, according to one of his schoolbooks. Truck drivers are okay, but loggers are not.

છે છે છે

At least two generations have grown up learning the environmental gospel from the popular public television program, *Sesame Street.* On March 9, 1994, *Sesame Street* showed an animated cartoon about a cruel woodcutter who chopped down the trees in a forest. Children glued to the television set could not miss the implication that cutting down a tree is an immoral act.

On a bitterly cold Chicago evening, a fashionably dressed woman leaving the theatre removed her mink coat, turned it inside out so only the lining showed, and carried it on her arm. "I'm freezing," she said, hugging herself for warmth.

"Why don't you put your coat on?" her husband replied. "I didn't pay a fortune for you to carry that coat." The woman pointed to a PETA group standing on the sidewalk, shouting, "Murderer!" and harassing other fur-clothed women.

"I don't know what PETA* is," the woman told her husband, "but they certainly don't like mink coats, and I don't want to be yelled at."

PETA — People for the Ethical Treatment of Animals

The President of the American Consulting Engineers Council complains: "We engineers suffer from the image that we make things run but don't get to run things. We must increase our visibility and enhance our image."

ೠ ೠ ೠ

"My wife says if she reads another lead paragraph in the newspaper that says 'Weyerhaeuser the timber giant,' she is going to scream. I tell her she is going to scream the rest of her life," Jack Creighton reported when he was president and chief executive officer of the Weyerhaeuser Company. (Scott Sonner, The Associated Press, December 19, 1994)

ೠ ೠ ೠ

The October 5, 1992, issue of *Chemical and Engineering News* carried a banner headline: "Chemical Makers Pin Hopes on Responsible Care to Improve Image." Frank Popoff, then-chairman of the Chemical Manufacturers' Association and Dow Chemical Company's Chairman and Chief Executive Officer, announced the unique program. "Responsible Care," he declared, "is the chemical industry's strategy for survival. It has earned us a place at the table in the environmental debate."

ೠ ೠ ೠ

"Lawyers hire agents to polish their image," reads a headline, explaining that lawyers are tired of "taking their lumps."

ೠ ೠ ೠ

Even an industry as benign as air cargo carriers lists as a priority: "Planning an aggressive public relations campaign to increase the image of and respect for the air cargo industry."

❦ ❦ ❦

And the irrigation industry, tired of being considered glorified gardeners with hoses, complains: "Want to watch blood boil? Ask what the biggest problem facing the irrigation industry is. They'll respond venomously, 'It's the public's perception of irrigation.'"

❦ ❦ ❦

Loggers, furriers, engineers, the forest industry, the chemical industry, lawyers, the air cargo industry, and irrigators all share the same problem: they're battling a poor image.

A bad image is a nightmare from which there is no awakening. The assurance that "sticks and stones will break my bones but names will never hurt me," so comforting in kindergarten, is a childhood myth. Maturity is learning that names will hurt. The desire for peer acceptance, as every parent learns, is a powerful motivator that impels teenagers to beg for brand-name clothes. It also motivates industries to spend large sums of money on public relations to improve their image.

Gaining a favorable image or a positive public perception occupies the attention of almost all organizations and industries today. The cigarette industry, the legal and medical professions, politicians, the government, and even inventors are worrying about their image and finding that changing their public image is more difficult than changing their products.

A negative public image damages credibility, erodes trust, and finally ends up limiting one's professional freedom to manage business or activities as deemed appropriate. The impact is far reaching because when the public perceives an industry negatively, activists have fertile soil for sprouting legislation to discipline and control that industry. A negative perception enables government agencies to command and control everything from trade to trees, pollution emissions from factories to secondhand smoke from cigarettes.

An organization stung by the slings and arrows of public criticism typically responds with a public relations campaign calculated to "educate" the public to understand its actions. Based on the assumption that when people don't know the facts they assume the worst, the organizations marshall facts calculated to win adherents to their causes. Elegant publicity packets, news releases and interviews, sometimes advertising, and the World Wide Web, telegraph the messages and facts calculated to win public acceptance.

However, when companies hit a catastrophe patchwork emergency actions may often prove futile. Louisiana-Pacific's fiasco was Oriented Strand Board siding coming apart on houses. They treated the problem with explanatory full-page newspaper ads, but the ads failed to stop the crippling lawsuits.

Philip Morris' full-page ads in the *Chicago Tribune* and other newspapers, accusing the media of publishing stories about the EPA's report on passive smoking that "exaggerated evidence," failed to sway the public.

Responding to biting criticism from the environmental community, the business world paints itself "green," and the green movement becomes a marketing tool.

Environmental organizations, liberal in their criticism of industry, are hardly immune to the sting of public

criticism themselves. Dr. Patrick Moore, one of the organizers of Greenpeace, pointed out in a 1995 interview that
eco-extremists are pressing for a new cause as the forest
industry gradually adopts the environmental agenda
Greenpeace promoted.

ANCHORED BY AN UNFAVORABLE IMAGE

An unfavorable corporate image is an anchor that
reflects industry's ties to its past rather than its future. Jack
Ward Thomas, chief of the United States Forest Service,
observed that the changing values mean "that what I was
taught early in my career; i.e., what is good for forestry is
good for everything else, has been rejected." Foresters are
unhappy that public concern for preserving natural forests
overrides the costs and benefits of forest management and
that decisions involve far more than dollars and cents. The
rank and file in the forest industry have been urged by
industry leaders to be "madder'n hell" about the closure of
the national forests. They may be "madder'n hell" about the
environmental victories that have limited the timber
industry's access to timber, but the environmental successes
have impaired traditional forest planning.

This remark of a lumber purchasing agent for a large
furniture manufacturer expresses the anger of forest
industry employees:

> "A lot of rich folks are out there throwing their
> weight around, environmentalists who don't know
> anything, making rules; before you know it there
> won't be any wood furniture because there won't be
> any trees.... Everything has gone to hell since Ralph
> Nader stuck his nose into things."

REPAIRING AN IMAGE

FEDERAL WAY, WA — The past decade has not been good to the image of the timber industry.

"I think it is going to be very tough to shake — the idea of timber barons, rape and pillage," said Jack Creighton, president and chief executive officer of the Weyerhaeuser Company.

By the year 2000, the industry will be more consolidated, with large individual holdings, bigger paper recycling operations and more international approaches to the marketplace, Creighton and several of his contemporaries predict.

But just as important as the visible changes in the makeup of their companies, they say, will be the public perception of their stewardship of the nation's forest resources.

"Our business is terribly visible," said Tom Ingham, president of the Simpson Timber Company in Shelton....

The public relations experts at the Plum Creek Timber Company in Seattle have been scrambling to remake the image of the former Burlington Northern subsidiary since it became an independent company in 1989.

That was about the time former Rep. Rod Chandler was quoted in the *Wall Street Journal* calling Plum Creek the 'Darth Vader' of the industry....

"It was our personal D-Day. It really shocked us," said David Crooker, director of timber operations for Plum Creek's Cascades region.

"Now we've taken the blinders off. It's absolutely necessary if you are going to talk about survival in the year 2000," Crooker said. "The days of thumbing your nose at the public and saying we are going to do it the way we want are over."

...Mike Bader, executive director of the Alliance for the Wild Rockies in Missoula, Montana, has been working for years to ban logging in national forests adjacent to Plum Creek land in Montana.

He said Plum Creek deserves some credit for changing logging practices "in some instances. But for the most part, I see those activities as just part of their advertising budget. Just PR for the masses. It's forestry with a smiley-face approach," Bader said.

Steve Rogel, president of Willamette Industries in Portland, Oregon, said the forest products industry was slow to respond to public criticism of its logging practices....

"We were behind the power curve on learning how to deal with public issues. My opinion is that we did get a black eye for that," Rogel said.

From ugly, square-mile clearcuts to corporate policies that kept the public in the dark, the timber giants have done much to earn their villainous reputation.

"Some people moved in here and bought large tracts of timber, borrowed money to do it, then logged off the lands to pay the debt. All of us are paying for that today," Rogel said.

Creighton and others agree that the industry's black eye is to a large extent self-inflicted.

"We did things thirty years ago we wouldn't do today. You can say that about most businesses, even life," he said. (Scott Sonner, The Associated Press, December 19, 1994)

Education, public relations, and lobbying have been the cures used by industries and institutions to remedy a troubled image. But, like Humpty Dumpty who fell from the wall in the familiar nursery rhyme, all the education and all the public relations have not been able to put Humpty Dumpty together again to restore a tarnished image.

Not that education, public relations, and lobbying are poor strategies. They do work and work well when they are in sync with public values. However, industries are finding that trying to convince the public to agree with policies that they have already rejected is doomed to failure. When the forest industry blithely explains the virtues of clearcutting, it simply confirms the public perception that loggers are villains with chainsaws. Clearcutting throws the industry into a lion's den, and it doesn't escape without injury.

The chemical industry, also suffering from a severely troubled image, is the flagbearer in the struggle to win approval by the public. When the alchemy expected from public relations and education failed, chemical manufacturers switched courses, spearheading a bold new program "to improve industry's environmental, safety, health performance and so regain public trust."

They called the new program Responsible Care.

THE RESPONSIBLE CARE MODEL

Some sources claim that the Chemical Manufacturers Association's (CMA) Responsible Care initiative can be compared to "trying to change the direction of a herd of mastodons rushing headlong toward a primeval tar pit and certain death."

The following excerpts from *Chemical and Engineering News,* May 29, 1995 illuminate the crisis:

> Dow Chemical's Frank Popoff says, "Responsible Care was designed to alter industry's performance, its commitments, and decision-making process." The operational word was "alter," not justify, industry's performance. The key to the program's success was improving "communications and understanding of and response to public concerns about products and plant operations." The

bold program benefited from the public advisory panel cracking the whip to keep the program in line with current public thinking.

Responsible Care principles were adopted in 1988 when the CMA's 175 member companies, whose 2,000 plants make almost 90 percent of United States' chemical products, responded to the changing American values. J.Robert Hirl, president and chief executive officer of Occidental Chemical, reveals the chemical industry's realization in the mid-1980s that gave birth to the program:

"All the while we were patting ourselves on the back for all the good we were doing, we ignored the shift taking place in what people were thinking about our industry. We thought people cared more about economic stability and jobs than they did about environmental matters. That was wrong."

Responsible Care pledges member companies will adhere to ten guiding principles that respond to community concerns, promote health, safety, and environmental considerations, and to operate facilities so as to protect the environment. The formidable performance objectives are backed by goals and six codes that identify nearly a hundred management practices that each member pledges to carry out. To implement the policies, a fifteen-member public advisory panel keeps a sharp watch over Responsible Care. Comments of some panelists reveal the magnitude of the challenge.

"PR-wise, the industry is somewhere between undertakers and used-car salesmen. Performance is what people look for, and if the industry makes an honest effort to be good citizens, then its credibility will increase." – Hoyt D. Gardner, former president of the American Medical Association

"Expectations are high in communities where Responsible Care has been talked about, but industry's lobbying efforts create the appearance of subterfuge." – J. Ross Vincent, chairman of the Rocky Mountain Chapter of the Sierra Club

Industry-organized and -managed, the voluntary Responsible Care program provides for members to assess themselves on compliance. However, self-assessment doesn't sell to some advisory panel members:

"To gain the credibility the industry wants or needs, it will have to go to independent assessments by an organization not beholden to the chemical industry." – Press J. Robinson, associate vice chancellor for academic affairs, Southwestern University

"Independent validation and, ultimately, community involvement are needed to make the program credible." – Mary Durkin, University of Chicago Lab School

Images change slowly. In May 1995, *Chemical and Engineering News* headlined a follow-up story on Responsible Care: "Responsible Care: Chemical Makers Still Counting on it to Improve Image." The story noted that the purpose of Responsible Care was to transform public perception from that of "an arrogant culture pursuing profits at any costs to one that could be trusted to protect public and worker health and the environment."

The story comments that the program reflects a clash of two cultures: the technical, and the value-sensitive public wanting trust, honesty, and credibility, *Chemical and Engineering News* reported, "Seven years into the program, Responsible Care and a $10 million per year advertising campaign have served only to stop the slide in the public's low opinion of the industry." Out of ten industries ranked, the chemical industry still stands [as of 1995] second to the lowest — the tobacco industry — in public esteem.

And there are still doubts in many circles. A member of the public advisory panel doubts that a cultural change has occurred in the industry: "When industry feels threatened, the wagons are circled and Responsible Care becomes irrelevant." The panel's insistence that industry advocacy and lobbying be compatible with Responsible Care principles and that only third-party verification of performance improvement will be credible have angered CMA. But they have also spurred new remedial actions.

Public approval of the chemical industry is slow in taking hold. A university-sponsored symposium on the use of pesticides and herbicides in a badly diseased forest repeatedly told the audience that chemicals were bad, but biological treatments were good! Responsible Care has not cured the sick image of the chemical industry, but the willingness of the Chemical Manufacturers Association to respond to the new environmental culture offers reason for hope.

THE FOREST INDUSTRY PURSUES A BETTER IMAGE

Like anxious parents taking the temperature of a sick child every thirty minutes, opinion researchers are frequently hired to poll the public's environmental temperature. Industry executives now recognize that the public's opinions about the environment and the forest industry are critically important for the industry's health.

The passage of the ill-famed "rider" to the salvage bill in 1995 inspired industries, environmentalists, special interests, major newspapers, television networks and individual stations to contract with noted pollsters to check anew the opinions of Americans about the environment. The polls showed a strong majority of the American public favors keeping and enlarging environmental protection and shows continued distrust of the forest industry's record on environmental issues.

The forest industry is on the trail of a better image, but to date the trail is twisting and turning, and the goal is elusive. The poll results are bad news for the forest industry. They indicate that the expensive campaigns to educate and inform the public have performed poorly in improving the public's trust in the industry:

Half of the respondents in a *Los Angeles Times* poll said the environment should be protected even if it costs jobs.

A Wirthlin poll found almost 75 percent of the respondents believe that protecting the environment is so important that continuing improvements must be made regardless of cost.

A Roper Poll for the *Times Mirror* asked participants if they thought that a logging company who wanted to harvest high-quality timber in its own forest should be able to do so even if the logging will harm a threatened bird. "No." replied 61 percent. The logging company should not be permitted to log if it will harm the bird.

Do You Like Trees?

When pollsters have asked substantive questions on environmental issues, the results have been informative: Americans' support for environmental reform is almost as strong as their concern for the environment itself. Yet it does not follow that they favor putting federal bureaucracies in charge of environmental concerns.

Most Americans support environmental protection, but they're not activists. Eighty percent of those surveyed consider themselves either "active environmentalists" or "concerned about the environment but not active."

A majority of Americans consider a political candidate's position on environmental issues when deciding how to vote: 81 percent rate environmental positions as greater than five on a ten-point scale of importance.

Most Americans support a greater role for state and local governments in environmental policy, and a reduced role for Washington.... Sixty-five percent of respondents believe that state or local government would do a better job than the federal government in dealing with environmental concerns, and 72 percent said that state and local government should determine what pollution control measures are used to protect air quality.

Most Americans support private property rights — they represent a "core value," in the words of Democratic pollster Celinda Lake — in the context of environmental protection. Indeed, poll after poll has found that two-thirds of Americans support compensation for landowners when environmental regulations restrict private land use in a manner that reduces property values.

The poll's findings confirm what many environmental reformers have long suspected: most Americans find nothing inherently anti-environment about regulatory reform, property rights, or devolution of power from Washington to the states. (*Wall Street Journal*, July 29, 1996)

The forest industry has learned that it must improve its image and is spending millions of dollars in the attempt.

Another measure of the importance of the environment is demonstrated by the changes in our language. Biological language now used to describe environmental concepts like "ecosystems," "biodiversity," "sustainability" trips off the tongues of today's schoolchildren and their parents. The word "biodiversity" is missing in a 2,000-page library dictionary published in 1966, and "ecology" and "sustainable" were defined only in narrow scientific terms. In the past, when forest management focused on timber and wood products, "sustainability" meant maintenance of the timber harvest. Not long ago, "biodiversity" made its way into the forester's vocabulary. Before the industry could get a grip on those concepts, ecosystem management acquired the aura of a noble pursuit.

However, it appears that ecosystem management means different things to different people. Nancy Davis, writing in *Nature Conservancy,* calls it "the Rorschach test of the 1990s. Everybody who looks at it sees what they want to see." Sustainability is the goal of ecosystem management, implying that we manage our forest lands with an eye to the future. (To complete the environmental trinity, biodiversity is a key component of ecosystem management.)

Though not clearly defined, these words sound "good" and express ideas that make people feel "good." Environmentally conscious landowners employ ecosystem management and incorporate biodiversity to assure sustainable management.

The forest industry has emulated the Chemical Industry's Responsible Care program with its own

Sustainable Forestry Initiative (SFI). Devised by the American Forest and Paper Association (AF&PA), an umbrella organization for the industry, the SFI has risked considerable internal dissension by embracing large parts of the environmental agenda. The Sustainable Forestry Initiative shifts from traditional forest management practices to the ecology concept, and moves from sustained yield to sustainable forest management, and ultimately designs new forest practices.

When they were formulating the Initiative, the AF&PA commissioned extensive survey research. Scott Wallinger, senior vice president, Westvaco Corporation, who chaired the Sustainable Forestry Task Force, describes some of the findings:

"The research showed that industry insiders believed that the negative perception was due to a communications problem — a lack of public knowledge. Outsiders believed that the industry did not have a communications problem, but rather a behavior problem that reduced its credibility with the public and legislators. The research commissioned confirmed the industry's poor image and also showed that sustainable forestry is a potentially powerful proof-of-performance. They also found that third-party involvement would show the industry's commitment to behavioral change."

The ambitious Sustainable Forestry Initiative was launched in 1995 and hit the street in 1996.

--·--

WHY INDUSTRY EFFORTS HAVE NOT IMPROVED THE IMAGE

The chemical industry is frustrated that its environmental progress has received so little recognition. The forest industry's effort to win public trust has not yet gone to trial. The chemical industry's efforts to win public trust and a more favorable image have had less than resounding successes. Those whose hearts have been in the program ask, "Why?" The answers range from "It's too soon," to "No matter what we say, the environmental groups are against us."

These answers avoid the real probing that is necessary in changing an image. Digging beneath the surface of an industry reveals possible reasons that account for the disparity between the intent and the outcome of the programs designed to change the public's perception:

1. The CMA's Responsible Care image-repair program is inner-directed. Both in language and concept, it addresses its own industry members and the environmental critics, not the general public.

2. The principles are based on actions that serve society, but do not specifically acknowledge a responsibility to society. Responsible Care's ten guiding principles "recognize and respond to community concerns." The Chemical Manufacturers' Association refrains from acknowledging that in their production of chemicals and stewardship of chemical products they are exercising a responsibility to society.

The public is more likely to believe an organization's sincerity when it affirms that it feels responsible to the public.

3. Responsible Care has not developed a method for gaining public support. A television campaign emphasized that a large number of chemical manufacturers are working together. The absence of a dramatic Earth Day-type activity enlisting the efforts of students and alliances of caring citizens has allowed the chemical program go unnoticed by the media.

4. Responsible Care is industry originated and industry controlled and has been criticized by the environmental community for not including third-party certification of their activities

5. The environmental community accuses and condemns the Chemical Manufacturers' Association for lobbying for actions that conflict with their stated Responsible Care principles.

6. The industry program is based on assumptions that:
    • Facts determine decisions and facts speak for themselves.
    • If we educate the public to understand the reasons for our practices, they will make "wise" decisions; i.e., decisions in the industry's favor.
    • When people are informed, reason will prevail.
    • Understanding leads to agreement.
    • Economic considerations guide the decisions of most people.

7. Failure to recognize that a sincere concern for the environment has penetrated into the consciousness of the American people.

Based on the Responsible Care evaluation, the SFI may also be in danger of not getting the results it seeks:

- The Sustainable Forestry Initiative affirms that it will meet the product needs of society. However, it resists the suggestion that, as the stewards of the forest, they also have a total responsibility to society.
- The Sustainable Forestry Initiative does not appear to have attempted to develop public support. There has been no forest industry equivalent to Earth Day.
- Wire service reports on the announcements of the Sustainable Forest Initiative began with how many millions the industry spent in a "new effort to log more responsibly and increase wildlife protection." There were no dramatic photos of earnest youngsters engaged in spectacular activities supporting the program appearing with the AF&PA releases. It would be wise to seriously question whether the public will be impressed by the millions spent by an assumed wealthy industry.
- At this time, the whole forest industry does not comprehend, understand, or embrace the Sustainable Forest Initiative (or programs other than specific trade programs that are primarily aimed at increasing sales and decreasing or improving regulations). Since the industry is divided into factions, these groups remain a barrier to total commitment and support.

- The forestry industry program is based on assumptions that:
  - Facts determine decisions and speak for themselves.
  - If we educate the public to understand the reasons for our practices, they will make "wise" decisions; i.e., decisions in the forest industry's favor. When people are informed, reason will prevail.
  - Understanding leads to agreement.
  - Economic considerations guide the decisions of most people.
  - The forest industry should avoid appealing to the emotions.
  - The public has to understand and realize that forest industry actions are basically good.

American Forest and Paper Association
By January 1, 1996
Compliance with the Sustainable Forest Principles and Implementation Guidelines will be "a condition of continued membership in AF&PA ."

Principles
- To practice sustainable forestry to meet the needs of the present without compromising the ability of future generations to meet their own needs. Practicing a land stewardship ethic which integrates the reforestation, managing, growing, nurturing, and harvesting of trees for useful products with the conservation of soil, air and water quality, wildlife and fish habitat, and aesthetics.
- To use in its own forests, and promote among other forest landowners sustainable forestry practices that are economically and environmentally responsible.

- To protect forests from wildfire, pests, disease, and other damaging agents in order to maintain long-term forest health and productivity.
- To manage its forests and lands of special significance (e.g., biologically, geographically or historically significant) in a manner that takes into account their unique qualities.
- To continuously improve the practice of forest management and also to monitor, measure and report the performance of our members in achieving our commitment to sustainable forestry.

Guidelines

- Broaden the practice of sustainable forestry by employing an array of scientifically, environmentally, and economically sound principles in the growth, harvest and use of forests.
- Promptly reforest harvested areas to ensure long-term forest productivity and conservation of forest resources.
- Protect the water quality in streams, lakes, and other waterbodies by establishing riparian protection measures based on soil type, terrain, vegetation, and other applicable factors, and by using EPA-approved Best Management Practices in all forest operations.
- Enhance the quality of wildlife habitat by developing and implementing measures that promote habitat diversity and the conservation of plant and animal populations found in forest communities.
- Minimize the visual impact by designing harvests to blend into the terrain, by restricting clearcut size and/or by using harvest methods, age classes, and judicious placement of harvest units to promote diversity in forest cover.
- Manage company lands of ecologic, geologic, or historic significance in a manner that accounts for their special qualities.

- Contribute to biodiversity by enhancing landscape diversity and providing an array of habitats.
- Continue to improve forest utilization to help ensure the most efficient use of forest resources.
- Continue the prudent use of forest chemicals to improve forest health and growth while protecting employees, neighbors, the public, and sensitive areas, including stream resources and adjacent lands.

THE AMERICAN PULP AND PAPER INDUSTRY'S PLAN

The American Pulp and Paper Institute (TAPPI) plan to cure the troubling public perception of the pulp and paper sector of the forest industry also rests on the Sustainable Forest Initiative (SFI) assumptions. TAPPI is attacking the public perception problem with the Paper Express, Operation Communications, and Earth Answers programs. TAPPI says: "By providing these programs to industry members, TAPPI is working to enhance our industry's image and attract students to our industry." It claims that these two goals are considered "crucial to the well-being of our industry." TAPPI launched the "Spreading the Word" program in 1995 in an effort to correct the public's misinformation about paper. Their aim is to provide trustworthy, unbiased, honest information to the public. Few would question the urgency of having believable, accurate information at the fingertips of TAPPI members. But will this factual information alter the public's perception of the industry? TAPPI believes it will. The communications manager of the program explains TAPPI's reasoning: "Since environmental issues are often emotionally charged, there's a lot of confusion about what is scientific fact and what is simply opinion or misunderstanding. TAPPI's position is that people should understand all sides of environmental issues before forming an opinion." (TAPPI, September 1995)

Common sense tells us that a major study is not required to recognize that rationality cannot compete with emotion. If one point of view is aimed at the heart, it is likely to prevail over one aimed at the intellect. The emotional approach usually carries the day. John Maguire was chief of the U.S. Forest Service during the first brushes between the environmental community and the forest industry. To frustrated complaints about the environmental tactics, Chief Maguire would reply laconically, "Emotions are a fact too." Appeals to the emotions put an uneasy burden on professional foresters and business administration graduates who manage the plants. They are not geared, either by training or nature, to using emotional appeals.

That the public does not weigh the most economical methods or the best management practices in deciding how much trust to invest in the industry is a bitter pill. Homeowners who suffer from the "sick house" syndrome from formaldehyde in particleboard feel no better because it is the most efficient and economical chemical for bonding resins. Homeowners consider only how they feel. Economy and efficiency are considered industry advantages, not the public's.

Western Wood Products and other trade associations explain wood's environmental advantages in their WoodWorks program. They believe that the comparative Lifecycle Analyses should give pause to anyone who thinks he can save the planet by substituting steel or plastic for wood. Although Woodworks has revealed stunning information, it has as yet failed to stun the public. Lifecycle Analysis is an intellectual approach that, again, tugs at no one's heartstrings.

The education programs, no matter how interesting and well conceived, compete with the passion and zeal of those who treasure the birds and forest mystique. A poignant letter from a timber worker, dislocated from his job by the virtual disappearance of timber harvested from federal forests, was printed in the trade journal *Timber/West* (April 1996). It expresses the feeling of many of his coworkers: "Americans just don't understand they've been brainwashed. Overcoming this by education may be too little, too late."

In an image contest between nature and industry, nature is a sure winner. Witness the advertisements of both the forest industry and environmental advocates who both feature the beauties of nature.

Finally, the forest industry is questioning the assumptions that guided their yesteryear efforts. New approaches offer more promise.

INTERNAL BARRIERS
TO IMPROVING PUBLIC PERCEPTION

Overcoming the internal barriers to improving the public impression of the forest industry is almost as difficult as keeping a cat calm on a hot tin roof. Formidable barriers resist almost any change:

Some forest industry executives refuse to heed what they consider as "every crackpot idea" from outside the industry.

Stakeholders have intimidated the industry with aggressive strategies that discourage action.

_._.—._.—._.—._.—._.—._.—._.—._.—._.—._.—._.—._.—._.—._.—._.—

Proposals for change breed internal dissension.

When the AF&PA imposed the Sustainable Forestry Initiative as a condition of membership, twenty-seven of the member companies are reported to have quit the association.

Within the industry, one faction bitterly insists that the industry has "sold out" to the environmentalists by adapting so much of their agenda. Other companies concluded that the industry had to do something to assure the public the industry is determined to protect the environment.

> WASHINGTON — Major U.S. wood products companies spent hundreds of millions of dollars last year putting into place environmental standards aimed at protecting timberlands, an industry trade group said Thursday....
>
> Although the short-term costs are high, industry executives said, companies needed to do something to assure the public they're determined to protect the environment.
>
> Some environmentalists praised the industry plan as a step in the right direction. Others, however, were more skeptical, noting that some of the proposals appeared vague.
>
> Other environmentalists had problems with the proposal and said it conflicts with the industry's position for continued logging in old-growth forests in the Northwest.
>
> A number of companies in the group are "actively bidding for old-growth sales," said Evan Hirsche of the National Audubon Society. Thursday's announcement is "a little difficult to reconcile with some of their actions on public lands," he said. (Bloomberg Business News, April 12, 1996)

Declining enrollment in university forestry classes and the merging of many forestry schools and departments into departments of natural resources signal a shaky foundation for the industry's future.

Some forest landowners deeply resent outsiders telling them how to manage their own land as an infringement of their private property rights. Loggers believe that ecosystem restrictions restrict their right to earn a living.

Many professional foresters dispute the recommendations of their associates. This is perhaps understandable since scientists disagree on interpretation of findings on the best management practices.

Trade associations answer plans to preserve forests by pointing out the greater environmental dangers of using other materials if less wood is available.

The industry resists proposals suspected of promoting world governance by the United Nations, centralized land-use planning, and landscape harvesting to minimize impacts on forest ecosystems.

They prefer to remind anyone who will listen that the greatest impacts on forests are agricultural land clearing, urbanization, eliminating fire as a management tool, and other activities.

Some industry members believe the forest industry should show leadership in concern for the environment. Others argue that this leadership isn't the industry's responsibility, nor should it come at the expense of taxpayers, rural communities, or the free enterprise system.

There is disagreement about the significance of public awareness:

"It is immaterial if the forest industry has a negative public image," say some industry leaders.

Others feel that public perception has a major impact on the fortunes of the industry.

Attitudes differ among regions.

The Western states, with more than half of their forests in federal, state and other public ownership, feel the public's concerns and the resulting regulations more keenly. The remaining major stands of old-growth timber found in the West are also rallying centers for activists.

The Southeastern states, with most of their forestland in industrial and private, non-industrial ownership, have long operated plantations. The Midwestern and forested Great Lakes states, with most ownership in non-industrial, private forestland, also feel the sting of public opinion. The more densely populated Northeastern forests have been cut over and are now being regenerated. Their unique concerns relate to forests that have been set aside for elite owners.

The arguments that plague the Western federally owned forests apply equally to the privately owned forests. The State of Minnesota has spent almost $5 million developing a Generic Environmental Impact Statement (GEIS) for Timber Harvesting and Forest Management. The GEIS outlined recommendations for improving forest management and harvesting practices, providing information, and increasing public understanding. Unfortunately, the effort was greeted with mixed enthusiasm.

The Minnesota Center for Environmental Advocacy answered that, though they've "identified degradation of resources … we're not getting them addressed.... We need to quit clearcutting to the water's edge.... We need to protect old-growth forests...." The Minnesota Ornithologists' Union said the GEIS recommendations merely passed on the tough decisions to a new bureaucracy.

Meanwhile, Minnesota environmentalist groups protested that cutting aspen trees in northern Minnesota forests at a faster rate than they can grow back is "geared toward producing timber to sell to mills and doesn't take into account how overcutting affects wildlife, soils, or tourism."

Attitudes differ depending on the major species of the regions. Hardwoods predominate in the East and softwoods in the West. As the hardwoods once imported from tropical countries become scarce, the temperate hardwoods increase in value.

Professional foresters resent that an uninformed public is attempting to dictate how forestry should be practiced.

The disconnect between segments of the forest industry prevents a coordinated top-to-bottom approach to any program to improve public opinion.

The forest industry, tired of being on the losing end of controversies, indulges in wishful thinking that the fire has gone out of the environmental movement. Charles W. Bingham, a widely respected industry leader, sees hope "in the movement of society toward collaboration, and away from confrontation." He points to "the partnership nature of many suppliers and customers" and "a more enlightened approach between labor and industry" and notes that, "Leadership at the community level is moving away from five or six good old boys sitting at a table making decisions." (Private communication April 30, 1996, to author.)

Environmental claims of deforestation are countered with statistics that show the improvement of the forests in the past sixty years using traditional practices.

The Pacific Northwest illustrates the hostility that develops when groups attempt to solve problems based on people's needs and perceived wants. The Bonneville Dam constructed on the Columbia River that divides Oregon from Washington has been providing relatively cheap electricity to the region for many years. Some scientists now believe that operating the Bonneville Dam is responsible for the virtual demise of the prized wild salmon, an endangered species. To ease the impact of the dam on salmon, the fish supporters propose a wind-generating facility. Ornithologists oppose the wind facility as a threat to endangered bird species. It's a typical dilemma: save endangered fish, hurt endangered birds.

The "biased" media

Ask the leaders and rank and file of the forest industry about the media coverage of their industry, and they quickly spit out: "The media is biased against the forest industry." Unfortunately, industry opinions about the media also color their feelings about public perception. Surveys show that the news media is not held in high esteem or seen as a credible source of information on complex and controversial technical issues. The media's low status has spurred the International Federation of Environmental Journalists to draft a set of principles for environmental and science reporters. One of the fifteen principles states: "Environmentalists are not necessarily right [no matter how dedicated they seem], and industry is not necessarily wrong [even though profit is a factor]."

Marketing a new image

Behavior psychologists tell us that once people have embraced a negative image, replacing it with a new image is extremely difficult. Examples abound: At this point in history, it would be very difficult to promote a positive image of the nuclear industry. On the other hand, it took many masterful campaigns to change the image of smoking from a sign of sophistication to a bad habit. As yet, no campaigns on the dangers of drugs have been effective in curbing drug abuse among youth. And still the efforts to change the image of chemicals, lawyers, and furriers haven't yet met with success.

Trying to sell intangible ideas using the same techniques for marketing products is a recipe for failure. Too, tactics for changing an image often butt heads with advocates for and against the proposition. It's important to remember that a negative image and the resulting confrontations and conflicts are cash cows for advocates of change. The accusation that environmental groups can't afford peace with industry appears to be borne out by the increase in membership during times of conflict.

Changing the public image of the forest industry is complicated by these difficulties. Marketing third-party certification, lifecycle analysis, and sustainable forestry practices involve changing the behavior of the forest industry and the industry's customers — clearly a daunting task.

Yet deep in the recesses of the marketing game lies an answer: idea marketing. Idea marketing enabled the American Forests organization to convince builders of single-family homes to save existing trees and plant new trees on the properties where they build. American Forests wanted to encourage builders to plant and save more trees, adding beauty to their properties and reducing pollution and energy costs. It was unlikely that builders could quickly recoup their investment by adding the costs to the sales prices of their properties.

How could they persuade builders to sacrifice time and money for these intangible future benefits? How could they change the behavior of builders and convince them of the value of putting dollars into trees?

Searching for answers to this tough challenge, American Forests recognized that the environmental movement had for the past thirty years wisely used idea marketing to change Americans' attitudes toward the environment. Using this strategy, the environmental consciousness-raising persuaded Americans to change their ideas about the earth, the forests, birds and wildlife. This idea marketing program also influenced Americans to save the planet by recycling, saving, and protecting all manner of things which had previously merited little attention. While the forest industry was busy success-fully marketing its tangible commercial products, the environmental community was successfully using idea marketing to persuade Americans to adopt a new archetype, new ideas, new values and, ultimately, new modes of behavior.

The American Forests organization could have tried to change how builders valued trees with traditional market methods like focusing on the value of trees. This technique would teach how tree growing will help the builders, use conventional advertisements, a slogan, and employ education techniques and public relations to deliver the message. These conventional strategies might have worked to some extent, perhaps increasing builders' tree saving and planting by 1 or 2 percent, a satisfactory shift of market share for many commercial products.

Instead, after studying the methods used to change the behavior of entire populations, American Forests opted to use idea marketing — a fairly new branch of marketing.

Idea marketing has been used to change people's behavior toward consuming energy, toward eating as affecting lifestyles that cause high blood pressure, and toward preserving rainforests. Idea marketing has revolutionized attitudes toward smoking, AIDS, and care of babies in underdeveloped countries.

Idea marketing — called social marketing in textbooks — uses the same techniques that market furniture, flooring and homes. But instead of altering buying behavior, it affects the voluntary behavior of target "customers" in order to uplift their personal welfare and improve society. The subtle differences between idea marketing and commercial marketing are evident in the idea marketing tactics used by American Forests. Idea marketing's objective is to actually benefit targeted populations or society, thus any benefits to the marketer are incidental.

American Forests sold the idea of trees as a benefit. A marketing-wise nursery could use commercial marketing techniques to sell trees (the product) to builders who have changed their behavior. Marketing a brand of toothpaste is commercial marketing. Marketing the practice of tooth brushing to a population that does not practice dental care is idea marketing. Idea marketing focuses on inducing the customer to take action, not just being aware of the product. In fact, the customer may not have to buy anything.

Implementing an idea marketing campaign challenges even the most experienced marketers. It requires far greater effort to fully understand the customers and the culture

influencing the customer's actions, as well as the ramifications of the behavior change and the barriers to change. Idea
marketing looks for partnerships with all types of organizations, such as schools, churches, and governments, that
might be involved for the change to be effected. The final
report of the social marketing consultant to American
Forests analyzed the research data and recommended three
strategies for messages to deliver to builders:
1. Focus on benefits to builders.
2. Show builders that saving and planting trees is a
    way to demonstrate innovative leadership to the
    community.
3. Persuade builders that they can easily plant and save
    trees, providing lectures, manuals, videotapes, and
    the like to inform builders how to take care of the
    trees.
Note the emphasis on the builders, not the trees. The
builders are persuaded and provided with the means to
change their behavior. The program succeeded in changing
the behavior of participating builders from careless damage
to trees during construction to appreciation of the value of
planting and saving trees.

Reinventing the Forest Industry bases its hope of
changing the public perception of the forest industry from
suspicion to trust on idea marketing. Idea (social) marketing
was invented by Americans in the 1970s and broke ground
marketing international health programs.

The forest industry itself is a target of idea marketing.
The many articles and advertisements urging people to use
blow dryers in wash rooms "to save trees," to use steel and
plastic instead of wood "to save trees," to avoid buying
furniture made from wood from tropical rainforests all

entrench the idea that saving trees is socially good because it means fewer trees will be logged.

The environmental community has co-opted manufacturers of steel, plastics, and other wood substitutes as partners in achieving a social goal. The emerging third-party certification of sustainable forest management represents an idea marketing effort to enhance the environmental credibility of the forest industry.

The aim of idea marketing is the good of society beyond merely delivering products for a better standard of living. Idea marketing does not preach a specific "truth" that everyone is expected to accept. From an Earth Force flyer published by the American Conservation Association, Inc., 1995:

> Team Up for Trees! Help Tropical Trees for a Healthy Planet! Kids are Taking Action for Tropical Rainforests. Some suggestions: Design tree greeting cards. Open a Team Up for Trees! store. Adopt community trees. Be a forest ranger. Set up a woodsy wishing well. Hold a tropical treats rainforest bake sale. Be creative and think of more ways to help trees. Ask a local business, bank, organization, or store to help you raise more money or match your pennies.

Marketing a new perception of the forest industry is a tough job compared to selling lumber or wood cabinets. It aims for a 100 percent market share.

The daily newspaper and television news programs hammer away at American's behavior with idea marketing. The campaigns by the national health organizations to raise awareness about cholesterol and high blood pressure is believed to have played an important role in reducing deaths from strokes by over 45 percent. The produce industry's "five a day" campaign has increased the consumption of fruits and vegetables. The National Oceanic and Atmospheric Association (NOAA) applied idea marketing to reduce marine pollution.

Even a casual study of environmental campaigns reveals how astutely using state-of-the-art techniques of idea marketing are promoting preservation of the environment. *Reinventing the Forest Industry* borrows their techniques and applies idea marketing to improve the public's opinion of the forest industry.

Chapter 5

Reinventing Forestry — Certifying Sustainable Forests

"Sustainability" rolls easily off the tongues of sawmillers and cabinetmakers, as well as Sierra Club and Audubon Club members, and educators and students. It's a comforting word that appeals to our longing for stability in a world changing too rapidly. A truce in the hostility between the forest industry and the environmental culture could be called based on agreeing that sustainable forests are a desirable goal. This means that the idea of embracing sustainable forestry is a significant step in moving the forest industry out of its position as whipping boy of the environmental culture to the dignity of credibility.

Though few experts agree precisely on what sustainable forestry means, the word sounds good to the public. Sustainability has come to signify the essence of the "green" movement worldwide. International conferences issue proclamations on sustainable forest management. The Seventh American Forest Congress held in 1996 advocated the concept of sustainable management in its Vision and Principles for American forests of the future.

Sustainability has achieved the status of a litmus test for concern about the environment. Sustainable forestry is now also a major issue for forest landowners and wood products processors worldwide. The forest industry is adopting sustainable forest management as the price of admission to the environmental political forum.

Sustainable forest management prescribes a full dose of environmental medicine. It requires foresters and loggers of these forests to simultaneously tend to water conservation, clean air, the ecosystem and biodiversity, as well as social values and economic impacts. If he had been confronted with the complex requirements of sustainability, Paul Bunyan, the legendary lumberjack hero, might have thrown away his axe and become a milkman.

Sustainability first jumped to the forefront when the growing concern about the depletion of the tropical rainforests hit the popular media. Sustainable forests represent an international concern that transcends national boundaries and cultures. Though accepting some of the international environmental definitions of sustainability makes the American forest industry uncomfortable, they agree that the value of sustainability is helping an international focus on the big picture in environmental discussions.

SUSTAINABLE FORESTS ARE "GOOD"

Everyone, it seems, wants sustainability, although sustainable development makes many in the environmental community nervous. Committed environmentalists view development as growth, the antithesis of preservation. Many oppose the concepts of the President's Council on

Sustainable Development, which builds on the vision of the pivotal Brundtland Commission's 1987 definition of sustainable development: "...to meet the needs of the present without compromising the ability of future generations to meet their own needs." The Fourth Forum of the World Chemical Industry, held in Paris in 1996, concluded that, "Sustainable development requires that economic, ecological, and social goals are pursued with equal vigor. This means that economic growth, efficient use of resources, and social stability are interdependent...."

Applied to forests, sustainability means managing forests "to meet the needs of the present without compromising the ability of future generations to meet their own needs." It combines equally reforestation, growing, nurturing, and harvesting of trees with conserving soil, air, and water quality, while maintaining plant and animal diversity — all in an aesthetically pleasing process. Sustainability challenges the forest industry to maintain beautiful forests and woodlands, pristine lakes, and pollution-free oceans, while providing necessities for a rapidly expanding population.

Will sustainable forestry assure that forests and woodlands, agricultural lands, water quality, streams, birds, plants, and wildlife remain intact over a long term, preferably forever? The environmental community and the forest industry both hope that the change to ecosystem management for sustainable forests implies that forest lands are managed with an eye to the future.

The concept of ecosystem management will profoundly impact every aspect of the forest products industry, far more than spotted owls or marbled murrelets, or assorted woodpeckers. Implementing ecosystem management has the "potential for wholesale change in how Americans view issues like property rights and the relationship of government to its citizens," according to Jack Petree, West Coast editor of *Pallet Enterprise,* a trade journal of the pallet industry.

The USDA Forest Service says ecosystem management "means...blending the needs of people and environmental values is such a way that [they] represent diverse, healthy, productive, and sustainable ecosystems." Ecosystem management will be used in land planning in which, the Forest Service says, "Boundaries [are] determined by ecological processes and structure, not those recorded in the county clerk's office." These are frightening words to private landowners who fear that the ecosystem's broad meaning will erase the "artifical" lines in the county clerk's office that mark their private property.

To beleaguered small landowners and foresters attempting to follow some guidelines, ecosystem management is a puzzle being invented and reinvented before they can fully grasp what it means.

MARKETING SUSTAINABLE FOREST MANAGEMENT

The forest industry is struggling to demonstrate to the public that it really intends to sustain forests and is not bent on "devastating the forests." Persuading the public that the forest industry means what it says challenges the most courageous realist. It's tough to overcome a lack of credibility. But, as Bob Hunt said in 1994, when he was president of the Western Wood Products Association, a major trade group, "The wood products business must respond to environmentalism in the market place or face shrinking markets and an uncertain timber supply."

The widespread interest in sustainable management reflects its menu of benefits.

The public has a vital interest in sustainable forests as essentials to life on this planet. Consumers concerned about the environment and quality of life benefit from the assurance that forests are managed to contribute to their welfare.

The environmental culture promotes sustainable management to further its values and, in its words, to "prevent the forest industry from trashing the forests."

The forest industry adopts sustainable management practices as its passport into the community of environmentally significant industries.

The forest industry, as stewards of the forests of the world, has a lively yet pragmatic relationship with sustainable management. Sustainable forestry assures the future of the industry. The forest industry also manages sustainability, paying the additional costs of ecological management. It is accountable to the public to preserve clean air and water and other benefits. Sustainable forest practices open doors to the world of respect and credibility. As a complex industry with a finger in almost every economic pie, the forest industry has many businesses rooting for its entry into sustainability.

A growing interest in the sustainable concept extends throughout the entire industry, from the landowner to the retailer. The retailer's interest in purchasing only from well-managed sustainable forests is explained by B&Q, one of Great Britain's largest chains: "The timber trade, and its connection with forest destruction, poses a challenge to any retailer of timber products who wishes to address the social and environmental concerns associated with this business."

Sustainable forest management is rapidly becoming a marketing attribute for wood products. Surveys conducted by Purdue and Pennsylvania State universities, the Institute of Sustainable Forestry, and others have revealed great interest, particularly among high-end customers, in purchasing products made from wood derived from sustainably managed forests. Third-party certification of sustainable forest management has become a feature that influences the buying behavior of environmentally aware consumers.

In recent surveys, 69 percent of consumers polled said they had boycotted or avoided a product due to environmental concerns; 72 percent felt that very few forested lands are being sustainably managed; and 76 percent said the forests of North America are under threat. Some surveys even show a willingness of environmentally concerned consumers to pay a premium for wood from these sustainable forests. Consumers describe "feeling good" about using products from sustainable forests. They translate certification to mean that the lumber or wood product did not originate from a clearcut forest, a rainforest, or an old-growth forest.

The forest industry hopes that the marketing attribute will motivate forest managers and producers to manage according to sustainable principles.

Agreement on the value of sustainability appears almost universal. In spite of this, methods of managing a forest for sustainability appear to have won only modest agreement. Environmental groups, still rankled by the memory of sustained-yield policies focusing solely on timber production, are suspicious about the forest industry's intentions. Many in the industry likewise are dubious that the money and effort spent on demonstrating sustainability will influence long-standing opposition.

As manufacturers of steel, concrete and other alternative building materials weaken lumber's strong hold on the residential construction market, forest industries are awakening to the realities of the successful environmental consciousness-raising in the past thirty or so years.

THE RULES OF THE GAME

Certification:
One Response to Environmentalism

"The wood products business must respond to environmentalism in the marketplace or face shrinking markets and an uncertain timber supply." That warning was voiced a year ago by Bob Hunt, president of the Western Wood Products Association....

One response by wood products producers has been a move toward the scientific certification of their forestry practices and products....

The AF&PA task force notes that many in the industry and the environmental community are skeptical about certification. Some in the industry view it as an effort to further restrict forest practices. Many foresters view certification as primarily a marketing tool, rather than a mechanism for demonstrating environmental sustainability. Environmentalists feel certification won't necessarily address all their concerns, and might be perceived as sanctioning unacceptable practices. (*Random Lengths,* "Yardstick, The Monthly Measure of Forest Products Statistics," June 1994)

How can the industry grapple with the skepticism about adopting new environmentally acceptable practices?

The industry is using two methods to gain admission into the environmental community:

1 Developing sustainable forestry principles that individual businesses must follow; and

2 Asking organizations that have won the public's confidence to vouch for the sustainable practices.

Sustainable forest management principles are featured in the Sustainable Forest Initiative (SFI) of the American Forest and Paper Association (AF&PA), which is improving forestry practices to create sustainable forests. The SFI exploits the favorable connection the public makes between sustainability and the environment. The language in the SFI is the environmental language now part of everyday vocabulary.

Subscribers to the SFI assess how they conform to the principles and report the results to the public. Lobbying activities that contradict the sustainable principles and these self-assessments are proving the weak links in the SFI (as they have in the Chemical Manufacturers' Association Responsible Care Model).

The SFI relies on companies to evaluate themselves, like grading their own exams. When the industry does not have credibility with the public, will self-assessment accomplish its goal? "No, it won't," say many members of the both the American and international forest industries and environmental advocates.

Instead, some sources believe public confidence can be restored only by following the sustainable guidelines created by an objective third party, and then paying them to judge and certify the company's sustainable forestry program. Only then are certified companies permitted to attach an eco-label to their products. Third-party certification is emerging as an international, industry-wide effort to demonstrate conformance to environmental criteria.

Third-party certifiers are emerging as a new segment of the forest industry, still tiptoeing on shaky ground to earn the support of the industry and the trust of the public. Independent certifiers evaluate producers and their output for forest sustainability, ecosystem maintenance, and social and economic benefits. These four parameters are signs of the 180-degree turnaround by the industry as environmental leaders.

Although industry has identified certification as one of the top issues of the day, doubts still haunt the process. Do consumers really care about certified wood products? Surveys show that some do, some don't. Are they willing to pay a premium for certified wood products? The answer: some high-end customers will; at this time, more won't.

Some traditional foresters look down their noses at certification as merely a marketing tool, a way to gain environmental credits. Dealing a worse blow to the integrity of certification are environmentalists who feel that certification permits the forest industry to cover up some unacceptable environmental practices.

The retailers, who face the realities of consumer behavior with every sale, are the pied pipers leading the industry to certification. They provide the marketing "pull." Home Depot, the largest American chain, early took the environmental lead in the United States by declaring that it is "committed to improving the environment by selling products that are manufactured, packaged, and labeled in an environmentally responsible manner." When a major retail outlet is interested in carrying certified products, it encourages forest landowners and manufacturers to budget the expense and additional processes for certification.

Certification grew out of the "green" movement that bloomed in the 1980s. "Green" has grown to mean a catchall for environmentally acceptable actions, such as acceptable harvesting methods, recycling, and packaging with "less is more." The Green Party has become a powerful political movement in Europe, and "green" has also come to represent the environmental movement in the United States. When a glut of green claims turned the "green" designation into an ineffective marketing tool, the industry searched for a more credible environmental symbol. Caught between recognizing the American dedication to environmental values and a nagging uncertainty that the green decade might be a memory that would fade in the future, the industry turned to certification as a solution.

> Manufacturers are finding it can be good business to be "green," to have it known their products are environmentally friendly.
> But who is to say what is what — how green is green?
> Although still relatively new, efforts to certify products on their environmental effects are building.... (*Professional Builder,* September 1996)

Late in 1993, the *Weekly Hardwood Review,* a publication for the hardwood industry, noted that, "The very beginnings of certification and accreditation issues are beginning to have an impact on the North American softwood industry, with hardwood likely to follow." Observing that some overseas companies were rapidly moving toward guarantees that the timber they use is harvested using guidelines on biodiversity, soil and water, aesthetics, and reforestation, the *Weekly Hardwood Review* questioned whether the expense could be justified in increased sales. The jury is still out on the increased sales, but the momentum gathering for certification for other purposes picked up speed by the mid-nineties.

> Environmentalists, who like hugging trees; and forestry firms, who like chopping them down, rarely get on well together. But a number of big Swedish firms, including AssiDoman and Stora, are now working in cooperation with green groups to develop timber "certification" schemes. Are these companies going soft, or just being clever?

> The idea behind certification schemes is to encourage consumers to buy products made from environmentally sound wood by giving such products a label or a stamp of approval. The tricky issue is to identify what sort of forestry (short of never chopping down trees) is environmentally sound. For instance, many environmentalists would like firms to reduce the size of clearcuts — areas of forest which are completely cleared — in order to protect habitats for wildlife. The Swedish firms (which own around 40 percent of the country's forests) are discussing what exactly they need to do to get their wood certified under the auspices of the Forest Stewardship Council, an international coalition of lobbyists and firms....

> In some European countries large retailers have promised to buy products from certified forests only. In Britain more than sixty big buyers of forest products ... have pledged not to buy uncertified products after 1999. These retailers appear to have decided that they will win more customers through green credentials than they will lose through higher prices. "The environment will become a competitive issue," says B&Q's Alan Knight. (*The Economist*, August 31, 1996)

Another type of assurance that companies are using wood from sustainably managed forests comes from the International Standards Organization's new ISO 14,000 environmental quality assurance procedures. The ISO

system is rapidly becoming an international trade requirement.

The International Standards Organization had previously developed a system of quality assurance called ISO
9000. The 9000 series deals with the quality of production
and is an accepted method used in international trade to
assure buyers of products that they are receiving the goods
they ordered. When a company signs up with ISO, it
prepares an elaborate report of its products, the standards to
which its production conforms, and the qualities the buyer
can be assured he receives. The company declares how it
will meet ISO standards, and trained ISO certifiers inspect
the plant and its products for conformance with the
standards it sets.

Until a couple of years ago, ISO concerned itself only
with the quality of products. But as the world gained
increased environmental awareness, and companies and
their customers sought assurance that processes and
products did not harm the environment, the international
community recognized that a change was necessary. To
accomplish this change, ISO appointed an international
technical committee to set up ISO environmental standards.
The committee, called TC207 (TC standing for Technical
Committee), after much deliberation devised environmental
standards in an ISO series called ISO 14,000. Though still
feeling its way, ISO 14,000 is rapidly becoming a *de facto*
international trade requirement.

TOO MANY COOKS

Worldwide interest in sustainable forest management
is gaining momentum like a loose rock rolling down a
mountain. Many NGOs (Non-Governmental Organizations)

and many governments want a voice in defining sustainable forests. The credibility of sustainable forest management is strained by too many cooks stirring the pot.

Early efforts focused on controlling the deforestation of tropical forests. Unacceptable practices in the temperate forests also received their share of attention. Two international organizations with different constituencies threw in their ideas on sustainable forest management:

- The International Tropical Trade Association (ITTO) is supported by the governments of forty-seven separate nations, including slightly over half of the countries that consume logs. The ITTO has established criteria for sustainable tropical forest management.
- The Worldwide Foundation (WWF) consists of twenty-two Do-It-Yourself (DIY) retailers and timber companies. The WWF has become the catalyst for certification schemes in Europe.

The interplay of the diverse interests has almost turned the sustainable forest movement into an international "soap opera," challenging observers to stay tuned to the daily maneuvers to learn who is involved with whom.

Attempts to force tropical timber companies into the sustainable mode range from pressuring buyers to stop using tropical timber to pushing importers to assure that wood originates from sustainable sources. Despite numerous conferences and rules, all the sustainable plans continue to suffer from lack of consensus on the definition of "sustainable management" and disagreement on whether forest management can really preserve biological diversity in tropical rainforests. Most environmental groups reject

logging in many tropical areas in order to preserve the traditional way of life of forest dwellers.

Around the world governments' responses to the poor
image of environmental standards varies. Britain responded
to the "environmental baddie" image in 1994 with BS7750,
a standard governing the impact a wood business has on the
environment. British wood businesses' evaluations are
mixed; some businesses consider BS7750 a good way of
defending a company's green record; others complain that it
is an expensive paper-chasing exercise. BS7750 hasn't sold
well in the environmental community either, because it is a
management system, not a performance standard of
required levels.

Time Is Running Out for the World's Forests

Trees are falling in the forests, and it is no longer a
question of people not hearing them. While the Pacific
Northwest debates the future of its forests, a similar
debate echoes through dozens of countries and hundreds
of communities around the world — with conservationists
right in the thick of it.

Groups like the World Wildlife Fund have spent the
better part of twenty years trying to reconcile the human
and economic needs of the developing world with the
preservation of biological diversity.

To claim that our overseas experiences should directly
inform the decisions made about the Pacific Northwest's
forests would plainly overstate the case. I do believe,
however, that our work in the developing world under-
scores just how complex decision-making processes can be
and how complex the thinking of conservationists must be
in response. (Kathryn S. Fuller, *The Oregonian*, June 10,
1996)

The momentum for international certification of
sustainable forestry reached a boiling point in 1993 and

boiled over into the United States by 1994. Many organizations have seized the opportunity in recent years to become the third-party certifiers. Each certifying organization devises its own criteria for judging the compliance of a forest.

Some industry decision-makers are skeptical that the preventative practices advocated by environmentalists will actually improve the environment. Despite some doubts about the value of the sustainability accreditation, it is fast emerging as an international forestry issue. Lively debates center on the relative merits of third-party certification versus self-assessment. The pro and con factions maintain firm positions. Mostly the industry and the public are still confused about the meaning of certification labels, if they are even aware of their existence.

Unfortunately, the move toward certification has not progressed as fast as expected, instead faltering in a sea of indecision and mixed messages. The industry has yet to get its act together on who's creating the rules, resulting in a confusing number of cooks stirring the certification broth.

In an industry as extensive, regionally distinct, and fragmented as the forest industry, consisting largely of independent entrepreneurs, many cooks may be expected. However, the existence of so many judges of environmental excellence creates a shaky base for credibility. Remember too that so many cooks confuse the public and dilute the effectiveness of certification in promoting stewardship of the forest and generally improving the industry's image. The situation cries for a chief chef.

Instead of suppliers, retailers, environmentalists, and end-users getting together to form a harmonious partnership to improve the present and benefit the future, potential allies are playing tug-of-war to position themselves as the authority on certification.

THIRD-PARTY CERTIFIERS

At present, the consumer who wants to purchase products made from wood that has originated in a sustainably managed forest finds confusion in the marketplace. There are a number of "eco-labels" that inform the buyer about the origin of the wood. The astute consumer puzzles whether the label is one of those "green" gimmicks or the real thing. Internationally, a consumer may be confronted by certification by Woodmark, a United Kingdom certifying program operated by the Soil Association; the SGS Programme, another United Kingdom operation; the Green Cross of the United States Scientific Certification Systems (SCS); and the SmartWood Program of the Rainforest Alliance in the United States.

At least ten organizations have emerged as third-party certifiers in the past few years. Spurred by the increased logging in the tropical rainforests, the World Wide Fund for Nature (WWF) pushed for the establishment of a system of independent timber certification to link timber producers with consumers worldwide. By 1993, the young certification movement reached the adolescent stage with the formation of the Forest Stewardship Council (FSC) to develop and assure consistent high standards. Even as the FSC struggled to obtain a footing in the certification arena, competing certification organizations announced their own agenda.

The FSC, the worldwide umbrella organization for certifying sustainable forestry, includes "socially beneficial" in its mission, in line with the environmental culture of responsibility to society. Third-party certifiers create their own principles and certify according to their own criteria.

An example of principles comes from the Rogue Institute for Ecology and Economy, a partner of the Smart-Wood Certifier, who, in turn, is certified by the Forest Stewardship Council:

1. Forest practices will maintain and/or restore the aesthetics, vitality, structure, and functioning of the natural processes of the forest ecosystem and its components.
2. Forest practices will maintain or restore surface and ground water quality and quantity, with special attention given to aquatic and riparian habitat.
3. Forest practices will maintain or restore natural processes of soil fertility, productivity and stability.
4. Forest practices will maintain or restore a natural balance and diversity of native species including flora, fauna, fungi and microbes, for the purposes of the long-term health of the ecosystems.
5. Forest practices will encourage a natural regeneration of native species to protect valuable gene pools.
6. Forest practices will not include the use of artificial chemical fertilizers or pesticides.
7. Forest practitioners will address the need for local employment and community well-being and respect for workers' rights.
8. Sites of archaeological, cultural and historical significance will be protected and will receive special consideration.
9. Forest practices executed under a Certified Forest Management Plan will be of the appropriate size, scale, time frame, and technology for the parcel, and

adopt the appropriate monitoring program, not only in order to avoid negative cumulative impacts but also to promote beneficial cumulative effects on the forest.

10. Ancient forests will be subject to a moratorium on commercial logging during which time the Institute will monitor research on the ramifications of management in these areas. (Rogue Institute for Ecology and Economy)

FSC member certifiers include the non-profit, New York-based "Smart-Wood," originally organized by the Rainforest Alliance. Another important certifying organization is the for-profit Scientific Certification Systems (SCS), of Oakland, California. FSC guidelines call for evaluation of a wood producer's environmental and social impact. SCS relies on independent panels of local people who live close to the evaluated forest, and uses scientists on the ground and people with an understanding of the specific site.

Although there appears to be mostly agreement with the principles but sharp variations in carrying it out, the whole muddle confirms the adage that "the devil is in the implementation."

Voluntary self-assessment is used by the International Standards Organization in its ISO 14,000. Conceived as an alternative approach to the international certification by the Forest Stewardship Council (FSC), the International Standards Association's ISO 14,000 generates uniform standards for assessing forest environmental management systems (EMS). Forest managers or companies can voluntarily adopt these standards, which certifiers from existing national standards organizations can verify. ISO 14,000 faces an uphill battle against fears that it directly conflicts with the FSC and the current lack of generic guidelines for environmental management systems. It has also run into a conflict with the AF&PA, who objected to the assumption that international environmental management standards for each sector are necessary.

Citing the voluntary nature of the ISO system, observers point to reducing duplication by a single set of international environmental standards and the resulting internal benefits, pollution prevention, and the possibility of lowering insurance rates. In fact, ISO 14,000 is becoming a *de facto* requirement for doing business in certain countries, and the growing global economy may leave individual businesses little choice but to obtain ISO registration. Some businesses believe that ISO registration will satisfy investors, the public, and environmental critics that their Environmental Management Systems (EMS) merit certification.

Voluntary systems like ISO 14,000 are becoming more popular with governments; in fact, so popular that governments worldwide are looking at adopting the system in their Environmental Management Systems (EMS) requirements. Some countries with strict environmental regulations see ISO 14,000 as a useful alternative to command and control regulations.

Skeptics question whether the ISO standards really lead to better environmental performance. Environmental advocates are deeply suspicions about self-assessment. The cost of qualifying may restrict the trading ability of small companies and may backfire in the role of nontariff trade barriers. A certain distrust of self-declarations and the fairness judging conformity adds to the unease about participating in the ISO 14,000 program.

Increasingly, concern that environmental certification may become a prerequisite for permits to market lumber is prompting companies to consider self-certification or utilization of ISO 14,000, the International Standards Organization's answer to environmental demands.

CERTIFICATION AND THE INTERNATIONAL MARKET PULL

The market pull for certification that lumber and wood products come from sustainably managed forests has a distinctly international flavor. Almost all major suppliers of wood products throughout the world are examining government and marketplace environmental concerns. It is too early in the certification efforts for clear answers about its value as an international market passport, or whether it may or may not command higher prices. The culture shift in the last few years, with everyone wanting to do the "right" thing (although they are not quite sure what the "right" thing is) indicates that suppliers are going to have little choice if they want to stay in the market.

> A beautiful, traditional Welsh stick-oak chair made by the Dyfed Wildlife Trust from timber extracted from the ancient forest of Pengelli, Dyfed. This woodland nature reserve is owned and managed by the Trust and the world's first "Woodmark" certification from the Soil Association. (leaflet from the Dyfed Wildlife Trust's Welsh Wildlife Centre)

A PRELUDE TO REINVENTING THE FOREST INDUSTRY

The AF&PA Sustainable Forestry Initiative and the third-party certification of sustainable forest management by the Forest Stewardship Council (FSC) members mark a breakthrough in forest industry efforts to earn public approval as a member of the environmental family. They are changing the industry to conform to public expectations instead of attempting to change the public to meet forest industry wishes.

Despite this change in direction, the industry is still seriously split on the value of third-party certification. Some landowners still question the wisdom of managing for sustainable growth. Others laugh at the entire idea of demonstrating that forests have been managed for sustainable growth and pooh-pooh third-party certification as a method of changing public opinion. Others believe just as strongly that certification is the path to winning those customers who care about the environment.

Some clear answers have risen to the top in the certification cauldron:

- Environmental certification does not substitute for product quality.
- Certification can be successful as a product differential.
- Certification does not substitute for innovation or productivity improvements.
- Most packaging producers feel that price and consistency have greater value than certification.
- Certification is more likely to attract a green premium in furniture, visible building products, and do-it-yourself projects.

ANOTHER STEP IN CIVILIZING FORESTRY

An historical perspective of the current certification excitement is that this is but the latest step in the civilization of forestry. It marks an era where forestry is a tool for civilization rather than merely subsistence.

Although Native Americans cherished the spiritual values of forests, they still used and managed them to serve their own survival needs. At first European settlers followed the pattern of using forests to meet their subsistence needs for food and energy, including removing them when they got in the way of planting fields. As Western civilization grabbed hungrily for more land, they used the seemingly inexhaustible forests to build cities, and cleared them to feed the growing populations. As the last century waned, finally worry about depleting the forests led to thinking and planning programs, policies, and procedures to make the forests sustainable.

Under the leadership of President Teddy Roosevelt, the National Forest Service cooperated with state and local governments, private forest owners, and other private-sector interests in devising policies and priorities to sustain the forests. These policies have guided forestry for the past 100 years. They led to suppressing and preventing fires, educating the public, establishing the profession of forestry, wildlife management, and other natural resource disciplines, forest regeneration, establishing tree nurseries, assisting forest landowners, reducing waste in making wood products, providing economic incentives for managing private forest lands, and establishing national forests to assure sustained timber production.

These policies worked. The United States was able to provide wood for homes and businesses for a population that continued to explode, without depleting our forests. We have about the same area of forests today as in 1920, and some areas, like the Northeast, have more as the U.S. agriculture changed from subsistence farming to commercial operations; it left the cleared forestland to regenerate, while transportation opened the way for the use of more fertile forest land.

As the population continues to expand and civilization spreads its wings, the cries for sustained forests grow louder. Forests grow, populations grow, and the pressures for certification of sustainable forestry practices will surely grow with them.

Chapter 6

Reinventing Wood and Paper

— · — · — · — · — · — · — · — · — · — · — · — · — · — · — · — · — · — · — · —

> The person who would forbid by law the harvesting of ripe trees from the forest should pass other laws to be consistent. He should make it illegal to sit on wooden chairs, to walk on wooden floors, to sleep in wooden beds, to eat from wooden tables, to rock his children in wooden cradles, and to accept shelter under a wooden roof upheld by wooden walls. When his daughter leaves his metallic home to attend her first junior prom on a concrete dance floor, he should make sure that she does not clothe her twinkling toes and graceful ankles in that part of a harvested tree called rayon. (Joseph T. Hazard, *Our Living Forests: The Story of Their Preservation and Multiple Use,* 1948)

Hazard captured the magic of transforming trees into products for shelter and comfort in his 1948 work, *Our Living Forests.* Forests provide for the physical and spiritual needs of humans. The production of wood products is where the rubber hits the road in the forest industry.

The reinvention of wood has been an unheralded contribution to the two-thirds of American families that now enjoy home ownership. In the face of higher timber and labor expenses, the forest industry has made durable and usable building products available at relatively modest costs. Reinvention requires reinvestment. For instance, factories needed to manufacture reinvented wood add up to astronomical stockholder investments beyond the reach of small businesses.

While the forest industry examines its environmental performance and revisits the way it relates to the world, wood miracles are making their way from research laboratories to the marketplace. As the poet said, "Only God can make a tree." But modern technology can create wood "better than nature."

It is, of course, still possible to sit on wooden chairs or walk on wooden floors, or to engage in any of the activities requiring wood. The incredible fact that these amenities are available today in quantities sufficient for the expanded population, despite thousands of years of harvesting the forests of the world, testifies to the major change taking place in the forest industry.

In previous eras when the industry exhausted the timber supply in one forest, it moved lock, stock, and barrel to the next available forest. The forest industry has a long history of surviving despite the brutal economic cycles that made or broke companies. When homebuilding slackened, they decreased production. When the housing market boomed, production boomed. Wood was plentiful; bark from the logs had little value. To dispose of sawdust and wood shavings and bark, they simply dumped the waste on their own mill property or landfills.

Today, there are few new forests where they can turn to for timber. Shortages have become the rule rather than the exception, and the public has turned its gaze on the environmental practices of the forest industry.

The only way wages can be paid and income provided to manage the forests is to harvest timber and other materials from the forest and then manufacture and sell these products. Without manufacturing forest products from timber nurtured over generations, the industry could not provide for the housing, furnishings, and other products to meet the public's needs.

Thus, industry is squeezed between the challenge of winning the approval of so many stakeholders, harvesting without devastating forests, and coping with a sharp decline in available timber. Truly a dilemma of heroic proportions! But with typical ingenuity and the zest for survival that characterizes the industry, it is solving the dilemma by reinventing wood.

Forestry gospel not so long ago held that a third of a log would be left in the forest, another third would be wasted in the mill, and the last third would become lumber. Today, all the factors controlling the forest industry have combined to reverse such wastefulness. The industry has reinvented wood to use the whole log, and the waste as the raw material for a new "species" that in some instances rivals Mother Nature.

Today, not a fiber is wasted; logs are completely put to use. Small pieces of wood glued together form composite panels or are finger-jointed for moldings. The bark of the tree makes animal litter and horticultural products. Any remaining scraps are turned into energy or valuable mulch. "Reused wood" is a fashionable substitute for new wood from forests. Urban wood — those decorative trees used in cities and towns — is harvested to make cabinets, furniture, and flooring.

Residues left over after converting logs to lumber, and small and low-grade logs once ended up in the sawmill teepee burner. Today they are reincarnated as engineered wood or composite boards. Composite panels replace solid wood in buildings, cabinets, furniture, or almost anyplace once held hostage by solid wood. Consumers rarely can detect the difference between solid wood and composites with a veneer* of solid wood.

The clear Douglas fir beams that used to grace homes have given way to new laminated beams. Polished furniture of walnut, black cherry, or curly maple has given way to veneered particleboard that resembles the solid wood. Customers can now walk into a store and select a table, a desk, or an entertainment center, all finished, ready for use as soon as they open the box and complete the assembly.

Bruce Plantz, the editor of *Furniture Design and Manufacturing* magazine, tells the story of friends who showed him their new "solid cherry" entertainment center. When Plantz told them it wasn't solid cherry but was made of particleboard with a fine cherry veneer, they were dismayed. They felt better when he explained that the composite particleboard was environmentally correct — preserving fine cherry trees by being made from the waste from mill operations and even recycled construction debris.

New types of wood forms and engineered beams to use in place of those that took nature 100 years or more to

---

*Veneer — A thin slice of solid wood glued over a composite board, almost indistinguishable from solid wood.

make are pouring out of processing plants. The industry has rejuvenated old technology to make paper products, achieved more extensive recycling than thought possible only a few years ago, and rediscovered tree species long ignored.

These new manmade species don't replace the beauty, warmth, and feel of solid wood; nothing can substitute for "real" solid wood. However, they do make it affordable for families to add that extra bedroom, or install handsome new kitchen cabinets, or a home office or home entertainment center without denuding another forest or a hard-earned, two-income-family budget.

ENGINEERED WOOD

The new wood species bear little resemblance to those that were familiar to the carpenter of a half-century ago. Chairs and cabinets, tables and houses use manmade wood designed especially to meet their requirements. When using solid wood, carpenters and manufacturers make do with the characteristics inherent in each species to build furniture and buildings. The "new" wood is created and tailored to meet the requirements of design.

Engineered wood is a wood composite designed to mimic the durability of solid wood while providing superior benefits of consistency and value. Engineered wood has acquired a generic meaning, like Kleenex™ and Xerox™, representing the general field of manmade composite panels. Composites — made of smaller pieces of wood glued together into a panel — have the smoothness and true flatness that welcomes paint, laminates, or wood veneers.

Engineered wood comes in an almost bewildering array of composite products to greet today's consumer —

all produced from wood scraps that used to be wasted. Carpenters and manufacturers could easily be intimidated by the new vocabulary of acronyms and trade names:

Particleboard
Waferboard
MDF (Medium Density Fiberboard)
OSB (Oriented Strand Board)
LVL (Laminated Veneer Lumber)
OSL (Oriented Strand Lumber; Scrimber,
    Timberstrand)
PSL (Parallel Strand Lumber; Parallam)
COM-PLY (Panels and Lumber)
Plywood
"Hi-Bond Slotwall"
"Multilook Prestige"
"CanPar"
"ECP" (an electrically conductive particleboard)
"Duraflake MR"
"Nova Appleply"
"Medite-FR" (flame-retardant MDF)

Wood fiber can be combined with plastics, cement, and other materials to make composites for special purposes.

Few of these products have reached the ripe age of ten years; most are newly invented to fill specific needs. Thinner particleboard is available for laminated flooring. Pre-finished hardwood plywood saves finishing time and costs, and special MDF makes superior molding.

These engineered or tailored products are eagerly grabbed by builders, furniture and cabinet makers, store fixture manufacturers, and other wood users. In the United States alone manufacturers are producing about twenty-one million tons of composite products every year — out of previously unused and unusable raw materials.

From an environmental view, engineered wood is the most environmentally friendly material available, a way of obtaining full potential from every tree harvested, using wood that would otherwise be wasted, minimizing the number of trees cut down. The famous ship the *Queen Mary* was known as the ship of beautiful woods. Writer T.W. Bousfield said that, "If the beautiful woods you see on the *Queen Mary* were solid right through, half the forests of the world would have had to be devastated." It is possible now to get up to 80 percent high-quality lumber yields because round logs aren't directly converted to rectangular boards. By combining pieces of wood, predictably sound pieces of lumber two-by-twelve-inch boards can be created from trees that could provide only two-by-four-inch boards through conventional methods.

From a woodworker's perspective, the properties of manmade wood are more consistent and predictable than natural solid wood. Manufacturers like particleboard and medium-density fiberboard (MDF) engineered or tailored for their specific products. Consumers find particleboard in bedroom and diningroom suites, kitchen cabinets and countertops, doors and shelves. MDF is often a substitute for solid wood in making quality furniture, tabletops, and side panels.

Engineered wood provides a sturdy base for the very thin slices of elegant woods called veneers. The veneers of today have little in common with those veneers of yesterday that had a propensity for peeling off.

Technology developed in 1960 made it possible to print a wood-grain veneer over the particleboard. It looks so much like veneer that few customers can detect the difference. They also appreciate the lower price points for furniture with a good appearance.

As might be expected with a new technology, the industry is agog with opinion, speculation, and recommendations. Comments in the trade journal, *Furniture Design and Manufacturing*, March 1996, from manufacturers who attended the 1995 Fall International Home Furnishings Market, show the industry ferment:

> Kathy Dickinson, president of Dickinson Designs, urges: "Quit finishing MDF so that it looks like imitation wood and capitalize on its value as a base for finishing. There's a huge opportunity for MDF to reign in its own arena, like with full finishing, *faux* and *trompe l'oeil.*"

> Office furnishings designer Thomas Perrin believes: "There are times when engineered wood is an excellent alternative to solid wood." Particle-board is cheaper than solid wood. And edgebanding and veneer add to its quality, making it just as good as solid wood. Particleboard is the only way to do larger products, because of the raw material shortage."

> A low-end retail furniture store owner, Amin Razak, declares: "Our customers are more interested in the fact that an MDF product costs less than a solid wood product."

> And Mary Badcock, of W.S. Badcock Corp., tells her customers: "We're not cutting down forests, and it's a cheaper product."

REINVENTING APPEARANCE

Homeowners like wood for its unique qualities, but mostly because it looks good and is naturally beautiful. Wood's good looks received a blow some years ago when a government pollution control regulation told woodworkers they couldn't finish their products with compounds that contained volatile organic compounds (VOCs). The blow turned out to be far from mortal. Now the industry uses waterborne coatings that can endow engineered wood with a rich finished look.

What's more, the high-quality, rapid-drying, high-solids waterborne coatings have been judged environmentally friendly (an industry goal repeated in advertisements). Combined with the environmentally friendly engineered wood, the waterborne coatings are helping the industry reinvent wood.

REINVENTING SPECIES

Not only has wood structure been reinvented, modern techniques are even reinventing the use of wood species. Alder, a West Coast hardwood, until very recently considered a "weed" to be eradicated, now appears in fine furniture often finished to imitate maple and cherry. Since 1982, oak has been the most popular species for furniture. Less-popular woods such as ash, elm, pecan, hickory, and hackberry are now finished to imitate oak.

Although maple is not as much in demand as it was in the 1960s, it remains in favor with many consumers; cherry is still a symbol of quality. Yellow poplar, gum, cotton-woods and basswood are frequently finished to look like

maple or cherry. Other less-expensive species are finished
to look like walnut and mahogany.

MAKING BIG PIECES OUT OF SMALL SCRAPS

As long lumber for molding, furniture frames,
flooring and wood components became more valuable when
it was available, the wood industry could no longer afford to
waste the short, random-length scraps resulting from
trimming boards to required lengths. The equipment to take
the scraps and "finger-joint" them into long, usable pieces
had been around since the late 1920s. However, it wasn't
until the timber shortages of the 1990s and the production
of faster, more efficient "finger-jointers" that finger-jointing
small scraps into usable lumber pieces took off. Now finger-
jointing operations are a way to gain environmental points
in saving trees.

REINVENTING USES

The ebullient energy of the wood products sector
extends to finding new uses for old wood. The wooden
pallet industry illustrates how what happens in the forest
influences other actions. The pallet industry, a major wood
business that provides the wooden pallets and containers to
ship everything from groceries to building materials, finds
itself in a stormy sea with rival captains struggling to either
maintain the status quo or change. The venerable National
Wood Pallet and Container Association represents pallet
maker interests. In this time of upheaval in all industries, the
question of changing their name and focus to be only the
Pallet Association, removing the "Wood" and "Container"

and incorporating materials other than wood, stirred passions on both sides.

The decision is more complex than appears at first. American temperate hardwoods are valued all over the world, even more so now when the world urges sparing tropical rainforests from the chainsaw. The catch is that most of our hardwoods are in the Midwest and Eastern parts of the United States, growing by natural regeneration after the pine and other softwoods were harvested. These "natural" hardwoods were not managed to produce timber, so a large portion are technically low grade, according to the rules of grading associations; many hardly pay for the cost of harvesting. The higher-grade hardwoods bring good returns, but if the landowners can't find a paying use for these low-grade woods, they can't afford to harvest them.

That's where the wood pallet industry comes in: A majority of wood pallets are made of these low-grade hardwoods, and they provide the market for those hardwoods that are not high enough quality for furniture, cabinets, floors or other products. The nearly 2,000 companies, 80 percent small businesses, and their employees, in the pallet industry see removing the wood designation from the Pallet Association as an opening for other materials. Some favor the change as a way of moving into the future; others fear the change as endangering their present.

To further complicate the issue, new technologies are making it possible to convert low-grade pallet stock to higher-grade furniture dimension — a development that keeps used pallets out of the landfills and may spare some hardwood trees.

The tale of the pallet illustrates why some who are reluctant to change are not necessarily "fuddy-duddies" but are faced with more complex issues than they are positioned to handle.

As early as 1934, the forest industry was challenged with the advent of competition from steel, plastics, and other materials. An editorial in the July 1934 issue of *Wood Products* asks, "Will the Sleeping Industry Wake in Time?" It asked if the wood products industry would keep its place in the markets because its products are being made from plastics, steel and other materials. No one is asking that question today about industry's new products and materials. Now the public is reviewing the use of all materials in order to use each one for its best qualities while conforming to the demands of our environmental culture.

Reinvented wood is giving the public good value — lower costs for products they need, cleaner air, elimination of waste, and recycling urban wood, used wood, and otherwise unusable fiber. Reinventing wood is part of the reinvention of the forest industry, a cradle-to-grave scenario in which growing trees and using trees are good for both people and the environment.

REINVENTING PAPER

Of all the miracles produced by the forest industry, reinventing paper could justify winning the grand prize. One of the saddest questions asked is, "How many forests did it take to make this Sunday paper?" Pulpwood forests are still bowing to the grim reaper, but that is changing rapidly. Today, almost 50 percent of the paper furnished is recycled from old papers.

Thousands of acres of land in the Willamette Valley that's now growing grass seed could be profitably converted to farming hybrid poplar trees for wood pulp and paper production, a new study suggests.

The softwood forests that were mowed for paper products are gradually being supplanted by fast-growing hardwoods that are ready for the paper mill in anywhere from three to ten years. They are harvested like crops, their new seedlings are planted, and in another few years they, too, are ready for pulp. Half the fibers that go into the pulping tank are now fast-growing hardwoods. Wood fiber is more than half of the cost in a pulp mill, and more mills are locating near large tree plantations.

Poplar Trees May Become Crop of Future

A scarcity of raw materials and rising prices for hardwood chips have made poplar plantations a more feasible alternative to grass seed than they have been in the past, the research concluded.

"Our production and yield analysis showed there could be a very positive return for poplar cultures at most sites in the valley," said David Hibbs, a hardwood silviculture specialist in the Department of Forest Science at Oregon State University.

Poplar grows best in well-drained soil with plenty of sand and gravel, Hibbs said, but the study demonstrated it also can grow in the heavier clay soils that often have been used in the Willamette Valley for grass seed.

Poplar could theoretically be grown for longer periods as a source of construction timber, Hibbs said. (OSU News Service, June 1996)

A more dramatic change is the movement toward chlorine-free pulping. The furor over dioxin and chlorinated organic pollution has led to new processes that take care of the environment in a holistic approach. These days each issue of the technical journals for the pulp and paper industry carries news of new developments in achieving environmental objectives.

The pulp and paper industry is acting on its belief that good environmental practice is good business. Another score for the environmental culture!

Chapter 7

The Reinvented Forest Industry

In the shadow of the spectacular Mendenhall Glacier, participants in a Management Institute listened intently as the speaker described the forest industry's new environmental stewardship.

A manager, brow furrowed, raised his hand: "If what you say is true, how come I haven't heard about these new environmental forest programs?"

The answer did not smooth his brow. "I haven't seen any change," he countered. "As far as I can tell, they're still clearcutting old growth."

. Interest replaced the frown after a brief explanation of the AF&PA Sustainable Forest Initiative and sustainable forestry certification programs.

But still the skeptic shook his head. "They say it, but I don't believe it. I don't trust the forest industry. It's still the same old exploiter with a facelift."

Others in the group nodded silent assent.

By virtue of its experience, knowledge, and education, the forest industry believes it should be steward of the forests of America. Instead, it finds that it has become a political eunuch, whose effectiveness in "meeting the needs of the present without compromising the ability of future generations to meet their needs" is undermined by a distrusting public. It is still pinned with the responsibility for managing the forests, but it has lost the authority to make many of the critical decisions affecting the future for the next generations.

The industry is awakening to the harsh reality that in today's environmental culture even the Congress that was once friendly to the forest industry does not have the political muscle to modify some of the environmental constraints.

In the 1994 Congressional elections, voters elected Congressional candidates who called the Environmental Protection Agency (EPA) "the Gestapo of government" and termed environmentalists "despicable." This happened in the face of public opinion polls that two-thirds of Americans call themselves environmentalists, and three-quarters of Americans complain that the government is not doing enough to protect the environment.

George Pettinico, a public opinion specialist, puzzled over this contradictory voter behavior in "The Public Opinion Paradox" (November-December 1995 issue of *Sierra,* the magazine of the Sierra Club). He ruefully suggested that Americans were more concerned today about urgent problems like drugs, crime, and health care than the environment. He concluded the article with the expectation that if Congress opposed environmental issues, it would eventually lead American voters to a stronger support of their environmental beliefs.

Managing both private and public forests is dictated more by public policy decisions, based upon the public's conviction that everyone is entitled to the environmental benefits of all forests, than scientific principles. The absence of boiling issues lulls some industry leaders into a false comfort zone, ignoring unresolved environmental issues coiled like rattlesnakes ready to strike, as these excerpts from a Sierra Club solicitation letter indicate:

> Dear Friend,
>
> As I write you, anti-environmental forces in Congress are escalating their all-out war on America's environment.
>
> If they succeed, they will rob us of our natural heritage — pollute our air and water ... cut down our forests ... close some of our beloved national parks ... and threaten the health and quality of life of thousands of American families.
>
> What is turning out to be the most anti-environmental Congress in history will — if we let them — literally wipe out some of the most vital protections environmentalists have fought for over the last 100 years!
>
> Ardent anti-conservationists are taking full advantage of their powerful positions in Congress — from which they are mounting a devastating assault on our nation's air, water, forests, parks and wilderness." (from the Sierra Club, Office of the Executive Director, fund solicitation letter, 1996)

Forests have a long life span; many species grow older than humans. In today's chaotic planning climate, long-term planning has become an oxymoron, one that the country can ill-afford now and will be paying for in the future. The rein-vention of the forest industry means paying the piper today so we can dance tomorrow.

With the change in values in society, caring for the earth is accepted as gospel, like the Ten Commandments. In the environmental culture that is now so ingrained in the American psyche, the forest industry cannot play its role of steward without gaining the public trust. The doomsday picture of a desolate, deforested world has had a frightening impact, and the public still holds onto the image of "the same old exploiter with a facelift." The only remaining avenue to winning public confidence is a sea change to transform the industry, to reinvent its mission and principles to blend with the current public beliefs about the environ-ment. *Reinventing the Forest Industry* is the sea change.

ON THE WAY TO REINVENTION

The forest industry is adopting full sustainable forestry, a goal shared by the environmental community. Sustainable forest practices are the beginning of the long journey to washing clean the images from the past that stain the forest industry's record.

The American Forest and Paper Association's (AF&PA) Sustainable Forestry Initiative, the emerging third-party certification of sustainable forestry, and the increasing interest in earning a position as a preserver, rather than destroyer, of the environment, demonstrate the industry's concern.

As an action plan to earn support of society, rein-venting is not an easy sell. Cassandras attract media attention more than Pollyannas. This applies to both sides, since warnings of environmental doom or the gloom of forest-dependent communities deprived of their livelihoods rouse passions and emotions more than the even-handed evaluations of industry. Actors in these conflicts over land, hunting, skiing, and camping are easily swayed by the environmental activists who throw their weight to further the causes.

The forest industry is already taking its first tentative steps, recognizing that it needs to "walk the talk." Typical of the cheerleaders for change is Professor Allan M. Springer, who wrote in the May 1995 Journal of *The American Pulp and Paper Institute* (known as TAPPI):

> The ... industry is entering a critical stage in its development in the environmental area. The low-visibility approach of the recent past has failed. The public has a poor image of the industry as polluting and reacting to environmental problems rather than leading in their solution....
>
> The ... industry needs a new comprehensive approach on how it views and solves environmental problems....

The forest industry is responding to the call by revamping old practices to demonstrate its sincerity in meeting the environmental challenge. The new coalitions, partnerships, and alliances marking the industry's turnaround would have been almost unthinkable a few years ago. It has finally taken to heart the comment of a long-time observer of the furniture business segment of the forest industry, W.W. "Jerry" Epperson Jr.: "As an industry we have resisted change, and in some respects have been slow to recognize opportunities that have been slapping us in the face. Our potential is being limited largely by our own credibility...."

Reinventing the Forest Industry is about transforming to a new customer-driven industry, with qualities that the public wants, and about marketing this new industry. This is not a facelift, not a rehash, or a sales gimmick, or a cover up. This reinvention represents an honest-to-goodness change. The forest industry is in the same position as an established restaurant that has served customers some poor meals, tarnishing its reputation. To stay in business, the restaurant has no choice but to create an entirely new cuisine, revamp the menu, and perhaps change the name and decor. The new menu before the public is an industry freshly tuned to the current American environmental culture.

A GREEN LIGHT FOR AN ENVIRONMENTALLY FRIENDLY FOREST INDUSTRY

Suppose the forest industry is not thought of as "destroying," "devastating," or "degrading" the forests? Will the public feel confident that the industry is acting in its behalf? Can the forest industry win the public trust? *Reinventing the Forest Industry* is based on the belief that the industry, by acting according to current standards, can earn the public's trust as a protector of the environment. The forest industry has already adopted many new initiatives that prevent the destruction, devastation, and degrading of the past, but it has not yet received the green light to manage its forests.

New commitment is needed to demonstrate that its practices link directly to the environmental, social, scientific, and economic values the public wants:

1. A commitment to *sustainable and environmentally
   appropriate* forest management.

Developers and public officials, naturalists and
foresters, environmentalists and the forest industry
support sustainable management, although the definition has an "Alice in Wonderland" quality,
"meaning," as Alice said, "what I say it means."
Like the environmental advocates, the forest
industry promotes sustainability to save forests.
Logs are the life breath of every wood products
manufacturer; without logs they're nothing. Without
forests there is no forest industry. Growing and
using forests for their best and highest use paves the
way to industry longevity. Downstream from the
forest, makers of flooring, furniture, building
materials, cabinets, and paper share the responsibility for sustainable principles put to use.

2. A commitment and a clear statement of its responsibility to society.

The forest industry feels abused, like the
comedian Rodney Dangerfield, who claims he "ain't
got no respect," for the jobs it provides, the communities it supports, and the quality products it
supplies. In the environmental culture, these social
benefits don't justify its old practices. The industry
needs to send a clear signal of its transition to
assuming responsibility within the current values of
society.

3. A commitment to science-based practices.

The industry cannot assume that everyone knows that wood is clean and green and good for the planet. Western Wood Products Association (WWPA) has joined forces with other associations to calculate the energy used in growing, processing, transporting, and using common building materials. Wood outshines steel, cement, bricks, plasterboard, and other common materials in the lifecycle race.

CONNECTING THE BACKBONE TO THE FOOTBONE

Another commitment the forest industry faces is to itself. The industry may be likened to an octopus that knows only his head and fails to realize how deeply his arms penetrate into American life. Though they love forests, most Americans feel little kinship with the forest industry or its links to their economic welfare. Neither the public nor the industry itself recognizes the interdependence of the many segments of the forest industry.

Foresters at the beginning and the retailers at the final end of the wood-use cycle share few of the concerns and information that affects them both. Sustainable forest management efforts are beginning to bridge that chasm and even make some strange bedfellows: today retailers unexpectedly care about forest management. The largest retail distributor in the United States, Home Depot, carries certified wood products (to be made of wood from certified sustainably managed forests), when they are available.

Retailers' money talks! The giant British distributor, Sainsbury's, with $2 billion in wood sales, has set the year

2000 as the target for buying only wood that comes from sustainable forests.

George White, of Sainsbury's, explains why they are going sustainable:

> Our customers expect Sainsbury's to know where its wood-based products come from. They also expect these products not to come from badly managed forests.... Certification is the route companies are taking to assure customers that their products come from sustainable forest practices.

Corporate buyers and environmental groups are now working together in the North American Buyers' Group for Certified Forest Products, committed to purchasing, or at least labeling, only certified wood products, as they do in some European countries.

In reinventing itself the forest industry will accomplish the goal of connecting the forest floor to the factory floor to the showroom floor.

A NEW LEAF FOR THE FOREST INDUSTRY

The forest industry is recognizing that to gain status in today's environmental culture it needs to turn over a new leaf, restructure its mission, and continue a shift to new practices and policies.

The new mission announces the intention of the industry to assume a new role in society, with responsibilities that stretch beyond providing wood-based products. It recognizes the complexity of forests and their special role in the well-being of the earth and its people; it recognizes that trees are everybody's business. The new mission statement

would be a credible demonstration that the forest industry
believes it is time for a change:

> Recognizing that forests uniquely provide for
> human and ecological needs, as steward of the
> forests, the forest industry assumes the responsi-
> bility to manage these resources for the benefit of
> society. The forest industry pledges to apply
> evolving knowledge of sustainable forest science
> and technology to meet the needs of the present
> without compromising the ability of future genera-
> tions to meet their needs.

THE SHIFT TO SOCIAL RESPONSIBILITY

These shifts in practices and policies would support
the idea that the industry has a change of heart.

- *Reinventing the Forest Industry* means taking a fresh
  look, operating in a new manner. The Statement of
  Intent is consistent with the increasing public
  pressure for industries to demonstrate a social
  conscience. It assures the nation that the industry is
  committed to change, to applying the lessons of the
  past to the future, and to walking its talk with
  actions that demonstrate stewardship. It shares a
  vision of the relationship of humans and nature,
  contributing to and depending on each other.
- There is understandable reluctance to expand the
  obligations to society, but closer examination
  reveals that the international forest industry started
  on this track around 1993 with the founding of the
  Forest Stewardship Council (FSC) — the main
  proponent of independent certification of perfor-

mance standards for forest management, and the major United States forest industry association of timber and paper companies, the American Forest and Paper Association (AF&PA):

> The FSC is an international organization of NGOs (Non-Governmental Organizations) dedicated to supporting environmentally appropriate, socially beneficial, and economically viable management of the world's forests. The ten principles include recognition of indigenous people's rights, workers' rights, and a "wide range of environmental and social benefits."

> The AF&PA, a United States association of the major wood and paper companies, whose members own about 13 percent of all forest land in the United States, in developing its landmark Sustainable Forestry Initiative (SFI), stated the belief that their member forest landowners "have an important stewardship responsibility and commitment to society." Their principles for sustainable forestry promise to "meet the needs of the present without compromising the ability of future generations to meet their own needs." They pledge the practice of an ethic that integrates growing and harvesting trees with the conservation of soil, air, and water quality, wildlife and fish habitat, and aesthetics.

• The shift to finding solutions

Reinvention means more efficient provision of the benefits of wood to the increasing world population. It proposes rational solutions to conflicting demands on the forests. It means using *more* renewable wood resources rather than substituting non-renewable steel, plastic, and cement materials. Wood, as a naturally renewable resource, helps stretch the supply of non-renewable resources for the future. It recognizes the continuing growth of scientific information and encourages the flexibility for change as research reveals nature's truths.

• The shift to practices appropriate for our environmental culture

    In the late 1960s and early 1970s, the way we viewed our relationship to the earth shifted to the new pattern that sired today's environmental culture. Now we are on the verge of another shift in the forest industry from practices keyed to an expansive economy to methods appropriate for our environmental culture. Many of the shifts are 180-degree turnarounds from past practices that have agitated the public.

    The nine new shifts are milestones in the history of the forest industry.

1. A shift from focusing on the welfare of the industry to include the welfare of the public, recognizing that the public has not bought into the proposition that what's good for the forest industry's bottom line is good for the public welfare.

2. A shift to a promise by the forest industry that, as the steward of the forest, it has the responsibility to nurture and sustain the forests with

consideration for all desired values.

3. A shift to promoting the understanding of the unique qualities of each tree species. The past practice of urging the public to plant trees as if all trees are equal gives the false impression that all trees have the same value. Such misinformation confuses the issues of natural and plantation forests, harvesting, and clearcutting.

4. A shift from managing forests as multi-value units to dividing forests into segments according to ecosystem attributes and a consumer-oriented classification of uses, including timber, endangered species, and recreation.

5. A shift from relying on "natural" forests for paper making to encourage plantations.

6. A shift from using ill-defined confusing buzzwords to clearly articulated forest management concepts that separate fact from theory.

7. A shift from an industry in which disconnected segments work isolated from each other, to a recognition of the interdependence of all sectors dependent on forests.

8. A shift from many systems of third-party certification of sustainable forest management, to an industry-wide coordinated program focused on market support for certification.

9. A shift from equating the objectives of "natural"

forests to plantation forests managed intensively to produce commodities. Genetic selection and other recent developments permit "designer" trees to be grown in environments that defy expectations from "natural" forests.

Instead of striving to maintain that elusive balance between environmental values and human needs for timber, these shifts attempt to solve some of the problems generated by differing values. Planning for some forests to remain truly "natural," for other forests to be managed to look and feel "natural," and a third category of forests to function as agricultural plantations can provide for all values. It clarifies utilization and reduces the insecurity shared by all forest stakeholders.

THE NEW FOREST INDUSTRY

The reinvention shifts, jump-started with the new mission, are supported by ten basic principles.

Reinventing the Forest Industry calls for two types of forest management: agroforestry and support of natural forests. The reinventing process is comparable to planting a crop: first you prepare the soil; then you put the appropriate seed and fertilizer in place; next, tend and nurture the seedlings and, finally, hope for favorable weather.

The soil has been prepared for *Reinventing the Forest Industry*, the seed has been planted by the AF&PA Sustainable Forest Initiative (representing industry), the principles adopted by the Seventh American Forest Congress (held in Washington, D.C., in February 1996), the

international Forest Stewardship Council (FSC) (representing environmental organizations), and the National
Forest Management Act of 1976 (NFMA) (representing
government). The proposed reinvention is the nurturing step
needed to satisfy the stakeholders in our environmental
culture.

How do we get from here to there? What will the
world look like when the forest industry has reinvented
itself? Will it be peace and harmony, with campers and
backpackers happily consorting with loggers? Will it be
rural communities enjoying stability of the economic
resource base? Will forests be managed to please the stakeholders and the stockholders?

The reinvented forest industry will not return forestry
to the "good old days," nor will the industry be under the
thumb of more regulations. The ten principles of
reinvention aim to provide a framework within which the
environment and the forest industry can live and work in
reasonable harmony. Following these principles will not
resurrect the Garden of Eden. Many a snake will offer many
an apple to disrupt planning and discussion.

The reinvented forest industry will fit neatly into our
current culture, providing a level playing field for the many
new decisions lurking in the background.

Each of the principles is arguable and will certainly
generate heated discussion. But, taken together, they will
improve the opportunities of Americans to enjoy and benefit
from their forests, both spiritually and materially, while
allowing the creations of nature to flourish.

PRINCIPLES OF REINVENTING THE FOREST INDUSTRY

Ten principles are articulated to facilitate the reinvention:

1. As the steward of a major renewable resource, the forest industry will nurture and sustain forests to meet human needs now and in the future for shelter, food, medicine, clean air and water, recreation, and other cherished values.

2. Forests serve many purposes, but each forest may not serve all purposes simultaneously. Each forest will be evaluated and managed for the most appropriate purpose compatible with the objectives of diverse ownerships.

3. Each forest is unique, but human activities can alter its characteristics. Forests will be managed and used to retain their unique characteristics.

4. The needs and fate of affected communities will be considered in planning for forest utilization, with sensitivity to the increase in conflicting demands for benefits from the forests as the population expands.

5. Natural forest land for development, agriculture, or other essential purposes will be utilized to the maximum extent feasible to sustain ecological systems and preserve unique characteristics of forest areas.

6. Agricultural forests (plantations) will be planned for the intensive production of fibers for timber, pulp

and other commodities. The best management practices (BMP), according to evolving standards, will be applied to intensive production and use of agricultural forests and plantations to produce timber and special forest products for commodity-type products.

7. All segments of the forest industry will be managed according to sustainable objectives to protect ecosystems and the economic, social, and health needs of the present, without compromising the ability of future generations to meet their needs and expectations.

8. The forest industry will study and recommend allocating public and private forested land for recreation, timber production, endangered species, and other emerging concerns, based on scientific findings. The allocation will be carried out by appropriate political processes, considering ecosystem and watershed preservation, rights of private ownership, and the needs and advice of local residents, landowners, recreation and environmental interests, and other stakeholders.

9. In allocating forests for specific uses, the consequences of industry actions on the health, safety, and amenities of life of the numerous stakeholders of the forest industry will be considered. Urban and occasional forests will be encouraged to help provide for the public's perceived entitlements for clean air, water, and aesthetically satisfying surroundings.

10. The interdependence and requirements of the many segments of the global forest industry will be considered in the location and operation of processing facilities.

Implementing these proposed Principles will not eliminate conflict. However, discussions of abstracts before the decisions directly affect participants should be less volatile and less confrontational, leading to more equitable solutions between the industry and society.

The reinvention program will deliver benefits to the forest industry beyond measure by bringing:

- A discernible improvement in the overall social and economic performance of the forest industry.
- Development of an identification system of forests as individual units (or zones or segments) and recommendations for allocation for specific purposes.
- Creation of a more favorable and supportive public image of the forest industry.
- A shift in using the industry's financial and energy resources from defending actions to proactively improving the industry's role in society.
- An outer-directed perspective in evaluating industry and government decisions and actions.

More detailed examination of these principles are in the following Goals and Management Guidelines:

Principle 1
As the steward of a major renewable resource, the forest industry will nurture and sustain forests to meet

human needs now and in the future for shelter, food, medicine, clean air and water, recreation, and other cherished values.

This principle will accomplish goals shared by the industry and environmental community by:

Providing information to the public on the diverse uses of forests.

Introducing a more comprehensive management concept of biodiversity.

Participating in simpler, coded, and clear environmental regulations affecting forests.

Organizing advisory panels to receive public input to the forest industry and to cooperatively solve emerging problems affecting the resource.

Empowering and strengthening the forest industry by joining the independent segments to accomplish mutual goals.

Discussion

This principle can be the platform for comprehensive public information on the reinvented forest industry. The goals may be accomplished by devising a consumer-oriented classification of diverse forests and uses, such as distinguishing between natural forests, regenerated forests, and forests designed for utilization as fiber for paper, solid wood, engineered composites, chips, and for special forest products. It would separate timber production forests from those forests that uniquely enhance bird, fish, animal, and human habitats, water quality, etc., and set apart forests for recreational purposes.

Using the theme that all trees have inherent value but all trees are not created equal, forests can be classified to encourage agroforestry in which forests are grown as crops for specific purposes. Intensive cultivation of plantations would relieve the pressure on remaining natural forests. (This is one of the Principles of the Forest Stewardship Council, the certification organization favored by the environmental community.)

Biologists and botanists value biodiversity for its contributions to the enjoyment of life, for economic reasons, as natural biochemical factories that provide medicines and foods, and for regulating local climates, absorbing pollution and carbon, and generating and maintaining soils. Special forests are consistent with the concept of setting up a system of "protected" areas and management in a regional context. Forests for special purposes will preserve selected organisms and species outside the natural areas.

If an ecological system is planned on a broad regional basis, specific areas could have efficiently managed and harvested monoculture zones and still fit in with the concept of ecosystem management. Some biologists are urging that endangered species be evaluated on a regional basis rather than on a species-by-species basis.

Participating in developing environmental regulations is important to the industry and the environment. "Compliance has become literally impossible," according to the Environmental Hazards Management Institute. The Institute reports the problem is acute for large companies and even more so for smaller ones because, "Regulations in the U.S. are so tortured, so patchwork, and sometimes so contradictory."

The forest industry does not know itself. Organizing a program that brings representatives of all segments together for a special purpose, such as improving their image by reinventing the industry, would enhance the strength and unity of the interdependent segments.

A charismatic industry leader with broad ties to several segments of the industry could propel the program forward and gain public attention. The chemical industry's "Responsible Care" program was created and promoted by leaders in the Chemical Manufacturers' Association and joined by other organizations in the industry. An organization within the forest industry, which would not be product specific, regionally specific, species specific, segment, or size specific could provide the basis for this program.

Another approach to leadership might come from joint action by professional organizations within the industry. Joint action by the Society of American Foresters (SAF), Forest Products Society (FPS), Society of Wood Science Technicians (SWST), the American Pulp and Paper Institute (TAPPI), National Association of Professional Forestry Schools and Colleges (NAPFSC), and similar associations would have impressive muscle. These partnerships may provide an easier consensus than the industry associations, but their endorsement would need to be ratified by industry if public confidence is to be significantly affected.

A public advisory panel could be the glue to hold the industry together.

In February 1996, some 1500 Americans interested in the fate of our forests convened the Seventh American Forest Congress in Washington, D.C., the first such meeting since 1976. Four days of intense discussion by foresters, landowners, environmental activists, and industry synthesized a vision and principles to guide American Forestry in the twenty-first century. These proposals for *Reinventing the Forest Industry* are consistent with the conclusions of the Forest Congress that advocated: "People's actions should ensure factual information and education concerning forests be readily available, engaging, and actively disseminated to all; science-based information should be accessible and understandable.

Principle 2
Forests serve many purposes, but each forest may not serve all purposes simultaneously.

Discussion
Controversy erupts when uses compete in forests, such as a ski lift on a desirable timber-growing site. Using this principle, the appropriate use of each forest zone can be determined. As an example, one forest zone within the total forest may be better set aside for organized recreation, another for diverse biological features, and yet another for timber harvesting.

The rights of individuals to use their property within the limits set by law is at issue. These rights are gradually being eroded by the provisions of the Endangered Species Act, wetlands requirements, local zoning, State Forest Practices Acts, and other considerations, creating uncertainty and financial problems for private owners, companies dependent on public timber, and the public's entitlements to pure air and water. This principle will bring resource stability to forest owners.

The Seventh Forest Congress promoted recognition and respect for: "All differences in goals and objectives of public, private, and tribal forest owners," and "the rights and responsibilities of forest owners." The Congress further concluded that: "Forest owners acknowledge that public interests [e.g., air, water, fish, and wildlife] exist on private lands and private interests [e.g., timber sales, recreation] exist on public lands."

Recognition that forests serve many purposes and that the unique characteristics of each forest zone may determine the best and highest use is key to the more rational, stable use of forests.

Principle 3

Each forest is unique, but human activities can alter its characteristics. Forests will be managed and used to retain their unique characteristics.

Discussion

When industry activities are known or suspected to have undesirable effects on ecosystems, human health or forest health, these consequences will be recognized and reduced. When natural forests are harvested, the timber removed may be replaced by planting appropriate species that preserve biodiversity. The capacity of the earth's ecosystems to sustain the resource varies from forest to forest.

The Seventh American Forest Congress urged: "Voluntary cooperation and coordination among individuals, landowners, communities, organizations, and governments is encouraged to achieve shared ecosystems goals."

Harvesting at the appropriate time to take advantage of the peak of a forest's life cycle will be sensitive to both human and ecological needs.

Each forest will be managed and maintained to preserve its uniqueness and health, diminish waste and reduce losses from desertification, erosion, and salinization. Management will use forests with greater efficiency than the current random-usage patterns.

The longleaf pine forests of the South demonstrate the outcome when management tinkers with natural characteristics of a forest. Longleafs at one time dominated the Southern forests, supporting an ecosystem with hundreds of plant and animal species, until commercial growers began to replace them with the fast-growing slash and loblolly pines. Now the loblolly struggles to survive in the sandy and rocky soils of longleaf country. Economic factors favor the loblolly, but nature is on the side of the longleaf.

The forest industry makes little effort to tell the public that some trees provide structural wood, others hardwood, or wood for unique purposes, still others mostly aesthetics, comfort, and wildlife habitat.

The suggestion to "save trees by not using paper towels," or to "save trees by not printing newsletters," telephoning instead (newsletter of the Institute of Deep Ecology, Winter 1994), and similar warnings teach that all trees have the same value. This, of course, flies in the face of the unique qualities of each tree species.

Principle 4
The needs and fate of affected communities will be considered in planning for forest usage, with sensitivity to the increase in conflicting demands for benefits from the forests as the population expands.

Discussion
The forest industry will plan programs to encourage management of forests according to accepted principles of sustainability. Programs such as the AF&PA Sustainable Forest Initiative to train loggers and private landowners in

sustainable practices will be expanded. Maintaining sustainable communities also involves having appropriate facilities to process, add value, and continue the chain of custody of products from the forests. Wood products with sustainable certification assure consumers that the manufacturing process has respected the rights and values of inhabitants and considered the community health, economics, and general welfare.

Regional planning based on the unique values of each forest will bring a new stability to communities. Landowners now subject to chaotic limitations on the use of their lands will be able to make long-term plans instead of swinging with the winds of change.

Voluntary, regional associations will be involved in planning for the highest, best use of each forest, accounting for the diverse objectives of publicly owned forests, commercial forests, and privately owned forests — particularly Non-Industrial Private Forests (NIPF). Incentive programs will encourage cooperation of individual landowners.

Each processing facility will bear responsibility for assuring that its practices protect the health, safety, and well being of the community according to the chain of custody of sustainable certification mandates. A community advisory committee, including local and national non-governmental organizations (NGOs), will give input to the facility on activities that may affect their health and safety, with scientists advising on solutions to identified problems. This procedure is supported by the finding of the Seventh Congress that: Forestry and management decisions must reflect the interdependence of diverse urban, suburban and rural communities.

Principle 5

Natural forest land for development, agriculture, or other essential purposes will be utilized to the maximum extent feasible to sustain ecological systems and preserve unique characteristics of forest areas.

Discussion

Land-use planning and Environmental Impact Statements consider the consequences of development on soil, air, water quality, aesthetics, and economic factors. Wildlife, fish habitat, special ecologic or geologic sites would also be considered.

This principle, which is supported by all the organizations concerned with sustainability and its certification, and the Seventh American Forest Congress, encourages planting appropriate trees in developing areas to replace or supplement revival of certain natural forest types. The species and sizes selected would include those suitable for preserving the natural ecosystem.

Principle 6

Agricultural forests (plantations) will be planned for the intensive production and use of fibers for timber, pulp, and other commodities. The best management practices (BMP), according to evolving standards, will be applied to intensive production and utilization of agricultural forests and plantations to produce commodity-type products.

Discussion

To implement this principle, foresters and forest products scientists would evaluate sites suitable for growing specific species for a family of uses and encourage new processing plants to locate near agricultural forests. This will make environmental sense for wood-producing plants, decreasing energy needs for transporting logs and eliminating duplicating infrastructure. The industry will produce credible, understandable fact sheets on the environmental benefits of manufacturing and using plantation produced products.

When it's possible, primary processing plants try to locate close to the timber supply from natural forests. When the timber is no longer available or accessible from these forests, the log haul increases, adding to the energy and cost burdens. As the harvest from natural forests is restricted, agricultural forests as sources of timber make economic sense.

Accessible natural forests with low-cost stumpage are the least-cost timber basket. Agricultural forest costs are becoming more competitive with natural forests as Forest Practices Acts, federal regulations, and public demands add substantial costs to harvesting from natural forests. By allocating some forests for commodity production, sustained-yield management for timber could be accomplished. This would also ensure long-term reproduction of key wildlife, and endangered species in other natural forests.

Implementing this principle could remedy the problem of sustained timber yield, which, according to the 1993 Society of American Foresters Report on Sustaining Long-Term Forest Health and Productivity, "...is seriously hindered by increasing regulation, appeals, or groups — often initiated by a few individuals or groups — with varying degrees of public support."

Tax incentives and land-use incentives for harvesting plantation forests could encourage voluntary compliance. States collectively have passed more than

250 measures, called green taxes, to promote environ-mental responsibility. Maine and Wisconsin use green taxes to encourage sustainable forest management on private lands. Other states provide incentives to encourage sustainable development.

A campaign to educate segments of the industry who work with buyers, builders, designers, and consumers would focus on the environmental benefits of using raw materials grown in plantations.

Principle 7

All segments of the forest industry will be managed according to sustainable objectives to protect existing ecosystems, and the economic, social, and health needs of the present, without compromising the ability of future generations to meet their needs and expectations.

Discussion

This principle advocates communicating to all segments of the forest industry and its stakeholders the goals and practices of the reinvented forestry industry, and describing the role of each segment in meeting society's forest and wood products needs. Trade journals, trade and professional associations, popular media, and educational institutions would carry the message to the entire industry.

Interdependency is now being considered when analyzing the technical and economic problems of constructing new processing facilities. The size and design of a sawmill depends upon the availability, species, characteristics, and volume of logs. The location of a furniture manufacturing plant considers the source of the lumber. Processing, distribution, and marketing are all changing to accommodate the requirements of sustainable management.

Principle 8

The forest industry will study and recommend allocating public and private forested land for recreation, timber production, endangered species, and other emerging concerns, based on scientific findings. The allocation will be carried out by appropriate political processes, considering ecosystem and watershed preservation, rights of private ownership, and the needs and advice of local residents, landowners, recreation and environmental interests, and other stakeholders.

Discussion

This principle grows out of the evaluations in Principle 2. Confrontations frequently occur over proposals for utilization in a forest, fostering piecemealing determined by the relative power of the many interests. The presence of an endangered species can shut down an entire public or private forest. Allocation would provide a blueprint for the diverse uses of the forest to accommodate all requirements. Presently, little attention is paid in local decisions to the ecology of the region and the economic interdependence of utilization decisions.

While allocations will most certainly generate heated discussion, this proposal assumes that decisions made in the abstract are more likely to be based on facts instead of the emotions of individuals involved in immediate utilization.

The allocation process would categorize and rate the social significance of the environmental values of forests, including such diverse benefits as:

Inherent environmental benefits: cleansing air, storing carbon dioxide, maintaining atmospheric quality, absorbing noise, absorbing pollution, protecting watersheds, purifying water, generating and maintaining soils, providing habitat for birds, animals, fish, insects, plants, and humans, preserving biodiversity, providing beauty and aesthetically pleasing vistas, providing climactic conditions for growing trees and understory products.

Utilization benefits, including providing timber for use in the shelter industries, fiber for the paper industries; creating opportunities for hunting, climbing, hiking, boating, swimming, skiing, camping; providing land for housing and resorts, and providing a wide variety of medicinals, foodstuffs, decoratives and other special forest products.

Using the forests for timber may diminish opportunities for utilization for recreation; either or both may impair the inherent environmental values of the forest. Historically, some values have taken precedence over all others, as in eras of expanding population when the forested land base is required for providing food or sites for homes.

In recent years, the unprecedented growth of forest-based recreation has added pressure for reserving land. The natural appeal of forests, plus their distance from crowded, stressful urban areas, lures many families to build homes in the woods, creating new challenges and conflicts over clearcutting, chemical use, and prescribed fire treatments.

When timber is harvested in recreational and forest residential areas, the recreation values may diminish. With increasing demands on a decreasing supply of timber from "natural" areas, the timber industry is forced to obtain logs where they are available.

Clearcutting timber in areas planted for the purpose of harvesting as agricultural crops or to preserve forest health, when performed with minimal visual damage would permit harvesting to minimize the impact.

Principle 9

In allocating forests for specific uses, the consequences of industry actions on the health, safety, and quality of life of the numerous stakeholders of the forest industry will be considered. Urban and occasional forests will be encouraged to help provide for the public's perceived entitlements for clean air, water, and aesthetically satisfying surroundings.

Discussion

The forest industry is now adopting the practice of re-engineering for pollution prevention rather than controlling pollution after the fact. The practice of environmental cost accounting (full cost accounting) will be considered in evaluating the effectiveness of the industry in this area. Many costs of pollution abatement hide in production costs. Separating pollution prevention and other hidden costs can provide significant opportunities for making decisions about production, environmental performance, and business planning for product marketing, use, and disposal, according to the World Resources Institute (WRI) "Green Ledgers: Case Studies in Corporate Environmental Accounting," 1995.

Publicly traded companies are already required by the Securities and Exchange Commission (SEC) to recognize and disclose certain environmental information in their financial reports. European companies are moving towards adopting the ISO 14,000 documentation of sustainable management practices as a marketing attribute. Another approach is applying existing indexes to measure environmental practices, similar to Arthur D. Little's "environmental performance index," giving information on environmental releases, regulatory compliance, resource consumption, and remedies. Proposed construction subjected to Public Acceptance Assessments or other measures that will impact public perception should decrease conflicts between the forest industry and the public.

Partnerships, similar to those formed by the Chemical Manufacturers Association, could expand these principles to include transportation and road construction, chemical use, waste disposal, and to the chemical companies and others serving the forest industry.

Principle 10
The interdependence and requirements of the many segments of the global forest industry will be considered in the operation of processing facilities.

Discussion
Increased competition for forest resources is to be expected in the looming era of scarcity, aggravated by the globalization of the industry. Domestic cooperation and coordination to a degree not previously experienced may become key to industry survival.

Where certain segments of the industry, such as harvesting special forest products, may be damaged by actions of other segments of the industry, like timber harvesting, simple notices to these companies in a timely manner to avoid deterioration of resources may be all that is required.

The interdependent forest industry segments will work together to achieve the goals of the industry. Certification of products from well-managed sustainable forests encourages cooperation and collaboration between industry segments. In the United Kingdom and the United States, where less than a handful of major retailers committed to buying wood only from certified well-managed forests, there was little consultation with landowners, lumber manufacturers, and wood processors about their role in making certified products available.

Brokers and wholesalers, lumberyards, processors, builders, architects, interior and furniture designers, and other segments of the industry are frequently on the outside looking in on decisions that affect them. They need to be included in discussions about the certification process to create both a market demand and products to fill that demand. The entire forest industry speaking to consumers with one voice can create the dramatic Earth Day type of demonstration that the environmentally aware forest industry deserves the public trust.

The Forest Industry Uprooted

When Macbeth was told that he would rule as long as Birnam Wood stayed put, he thought himself safe. Shortly afterwards he was vanquished by an army from the south that had advanced behind branches chopped from Birnam's trees. The men who run the great forest companies of North America and Western Europe should perhaps brush up on their Shakespeare. They too are threatened by invading foresters from the south.

For more than a century, the woods of the northern hemisphere have supplied most of the world's forest products. Forestry companies in North America and Europe have had two advantages over their southern competitors: vast forests that are close to the world's biggest markets, and a ready supply of capital to pay for the capital-intensive business of turning trees into paper. But they are also facing a growing handicap, in the shape of forest-loving environmentalists. And the South is making better use of an advantage of its own: a climate in which trees grow faster.

In contrast, companies outside North America and Western Europe seem better placed to respond to the growth in world demand. Russia's resources are vast — it contains more than a fifth of the world's forests — although most of its trees are far away from major markets. Asian and Latin American forests are closer to areas of fast-growing demand. Local companies such as Indonesia's APP and April, two family-controlled firms, have met much of the region's growth in demand for forest products. In Brazil, Aracruz, a highly profitable forestry firm, plans to expand capacity by a fifth over the next two years and has been selling increasing amounts of forest products to North America. (*The Economist,* August 31, 1996, pp. 3-54)

Chapter 8

Which Way for the
Forest Industry?

The life of every American, the destiny of every community, and the fate of every level of government is linked in some way with the forest industry. Americans worry about their forests, and many feel that the forest industry is not caring for them, according to the new standards of accountability for the earth.

In the past thirty or so years the country has changed from faith in unlimited abundance to fear of living on a fragile planet with limited resources. The environmental culture is now part of the American psyche and the root of current attitudes about the forest industry.

The forest industry has borne a large portion of this cultural change and a disproportionate share of the environmental community's scorn. The industry is recognizing with mixed enthusiasm that to be outside the environmental culture is a barrier to carrying out its responsibilities as steward of the forest.

With this recognition has come a determination to climb over the barbed-wire fence and join the environmental community in order to meet the needs of today's world, while providing for future generations to meet their needs. The forest industry has taken bold steps to become part of the environmental culture, rather than its adversary.

The cornerstone of saving the forests is sustainable forestry management. The major industries in the American Forest and Paper Association (AF&PA) have embraced this mission. The Forest Stewardship Council supports the sustainable forestry concept. Federal and state governments have adopted the sustainable forestry goal as the core of new regulations.

Sure, there are many details to be worked out, but sustainable forestry is a breakthrough. Gaps still exist in the commitment, and these gaps form another hurdle for the industry before it is warmly welcomed by the public as allies in caring for the earth. Filling these gaps is the objective of *Reinventing the Forest Industry.*

There remain large pockets of resistors to any partnership with environmentalists. These fist-clenched, fight-back members of the industry are hoping that the environmental community will "wake up to reality," and the public will finally "see the light" before they themselves have to change.

The environmental community is changing, but the light it sees is not the same one that fills industry's dreams. The reality is that the zealous baby boomers who created the first Earth Day are waking up, not to the side of the forest industry, but to the realities of their own success. Having invested their hearts and souls in saving the world, they are now reaching middle age and finding that the powerful environmental organizations they created must worry about meeting payrolls and paying expenses, just like industry. The future of the earth often must be pushed to the back burner.

There are still diehard environmental advocates and many non-governmental organizations (NGOs) that have a stake in baiting the forest industry. They are ready to jump on any bandwagon to protest for their cause.

So both sides of the bell curve of opinion are committed to continued animosity. But mainstreamers in the environmental community, like those in the forest industry, recognize that the world faces too many problems to allow the luxury of bickering. They admit their insensitivity to farmers and loggers and others who have been deprived of a livelihood by the environmental ardor that swept the country

With more *give* on the part of both the forest industry and the environmental culture, negotiation instead of litigation, working together rather than against each other, becomes the objective. The reality is that many of the disputes have little to do with the environment and much to do with so many people vying for the same space.

The problems plaguing the world will increase in the next century. The world can no longer afford to have two groups like the environmental community and the forest industry at each other's throats.

Will the forest industry drown in arguments on the details of reinventing itself, or walk the next mile to become part of the environmental culture? Will the environmental community stand firmly on its old party line, or open its arms to the efforts of the forest industry?

Without both sides shaking hands, public cynicism will not convert to public trust.

Without the public trust, the forest industry will not be positioned to do its job in caring for the forests.

Quo Vadis for the forest industry? The situation cries for leadership in collaboration and cooperation. Life on earth can no longer tolerate so many gladiators!

Index

To order additional copies of

Reinventing
the
Forest Industry

Book: $16.95   Shipping/Handling $3.50

Contact:

BookPartners, Inc.
P. O. Box 922
Wilsonville, Oregon 97070

Fax: 503-682-8684
Phone: 503-682-9821
Phone: 800-895-7323